THE OUTLOOK FOR PULP AND PAPER TO 1995

Projections of worldwide
demand and supply

Implications for capacity and world trade

FOOD AND AGRICULTURE ORGANIZATION OF THE UNITED NATIONS
Rome, 1986

M-38
ISBN 92-5-102433-2

Foreword

Paper is an essential material for the growth and development of society and its economy. The main raw material in paper manufacture is wood, and the industry consumes more than one-third of the world's annual production of industrial roundwood. The development of the paper industry is important to ensure the rising standards of communication and material well-being of people everywhere. The efficient growth of the industry contributes to employment in the forest, in mills and in the processing industries that utilise the industry's products. To secure this growth there must be appropriate investment in the industry, in the forest, and in the infrastructure and training of personnel which support these activities.

This study provides a perspective of the development of the pulp and paper sector to 1995. It is thus the principal subject for discussion at the Expert Consultation on Pulp and Paper Supply and Demand Outlook at FAO in Rome on 16 to 18 September 1986. The report from this meeting discusses action to secure the optimum development of the sector internationally.

This study has been prepared under the Programme of Outlook Studies of the Forestry Department. Central to the study are econometric projections of demand and supply and the consumption of inputs. These draw on the international data on forest products developed in FAO. Priority for this study was recommended by the FAO Advisory Committee of Experts on Pulp and Paper. A trust fund was established to support the work, and it was supported by the pulp and paper industry of many countries. This has permitted the mobilization of consultants to assist on a number of specific topics. The study was further assisted by the participation of industry experts in an Industry Working Party which provided information and technical advice on every aspect of the study. Those who attended meetings or carried out specific work are listed in an annex. This opportunity is taken to express sincere appreciation to all those who have contributed to this major international cooperative effort.

M.A. Flores Rodas
Assistant Director-General
Forestry Department

CONTENTS

GLOSSARY OF TERMS AND ABBREVIATIONS

UNITS OF WEIGHT:

tonne(s) - metric tonne(s)	mt
kilogram(s)	kg

UNITS OF VOLUME:

cubic metre(s)	m^3

UNITS OF VALUE:

United States dollar	US$

MAGNITUDES AND RATIOS:

thousands	1 000
millions	1 000 000
thousand millions	1 000 000 000
grams per square meter	g/m^2

MAIN ABBREVIATIONS:

Gross domestic product	GDP
Gross fixed capital formation	GFCF
Industry Working Party	IWP

INTERNATIONAL ORGANIZATIONS:

Organisation of Economic Co-operation and Development	OECD
United Nations	UN
United Nations Conference on Trade and Development	UNCTAD

GROUPS OF COUNTRIES:

World
Developed market economies
Developing market economies
Centrally planned economies

Developed countries include *developed market economies*, USSR and Eastern Europe
Developing countries include *developing market economies* and Asia *centrally planned economies*

The composition of each group is listed in ANNEX I, and is illustrated in the Figure.

CHAPTER 1

INTRODUCTION

Good and reliable international statistics are essential for planning viable pulp and paper industry development worldwide. This special *FAO Outlook Study* has been carried out in close cooperation with the pulp and paper industry to provide meaningful projections for worldwide demand, supply and trade of pulp and paper products to 1995.

The *FAO Outlook Study* has employed the latest international statistics, innovate econometric methods and specially developed computer models for demand, supply and production capacity. They have been designed to be updated for future work. This work benefits from over three decades of FAO experience in this area as well as from the active contributions by more than 130 experts from 43 countries.

That paper is a commodity vital to the growth and development of every country, its communications and packaging, is beyond dispute. The worldwide production of paper and paperboard in 1984 — 187 million tonnes, valued at US$100 thousand million — represented fully 1 percent of the world's total economic activity.

The international character of paper — both its production and consumption — is clearly reflected by the fact that 90 countries produce paper and virtually every country in the world consumes it. Some 150 countries import paper and paperboard today, and in 1984 39 million tonnes of paper and paperboard were traded internationally plus 21 million tonnes of pulp. This trade was valued at US$30 thousand million contributing some 1.5 percent of world merchandise exports.

Recognition of the pulp and paper industry's importance in the forestry and forest industry sectors (for wood is its major raw material) in meeting social needs, the international interrelationships involved and the long-term implications of decision-making has lead to the priority given to outlook studies for this industry.

How the *Outlook Study* was carried out

The broad objective of the study was to establish a uniform, unbiased and international view of requirements, and an international perspective of the sector's long-term development. This should provide a reference framework for policy formation and strategic decision-making for the sector and its participants.

This *FAO Outlook Study* is based on computer models using the extensive FAO historical data base and other international data. Two different economic forecasts for the development of the world economic scene to 1995 have been used, both for comparison purposes and as alternatives. One is based on FAO and United Nations sources (called the *FAO scenario*), while the other is from the *CHASE Econometrics 1986 Long-Term Report* (called the *CHASE scenario*).

Important assistance throughout the course of this study has been provided by an Industry Working Party (IWP), made up of representative experts from the paper industries of many countries. The structure and assumptions of the various models as well as the resulting demand projections were carefully reviewed at numerous meetings with the IWP. In addition, the IWP has advanced views on changes in end-use patterns for paper products and competing products, and has helped in areas where FAO data were not available.

About 130 industry experts from 30 countries contributed directly to the study through participation in the FAO Advisory Committee of Experts on Pulp and Paper, its Subgroup and the IWP (ANNEX II). Additional information was volunteered from 13 other countries. Many more worked within various countries to support this contribution. Four sessions of the Advisory Committee and seven sessions of its Subgroup considered the progress of the study. The IWP held 17 sessions in 11 countries: five were held at FAO in Rome, 12 were hosted by the industry (FIGURE 1). In addition, industry experts worked in special sessions on four product areas (ANNEX III).

The aim of these meetings was to identify factors of the real world, inadequately represented by the econometric models, which may be expected to influence the course of development. It was also possible for some countries to draw on the much more detailed information on inputs, production processes, product composition and end-use. In particular, changes in technology and in end-use, and developments in competing products were examined to identify trends which influence consumption of paper and its competitiveness. Thus it was possible to obtain a fuller description of the sector and

2

Figure 1

the perceptions of experts of its performance, though in a qualitative way.

The main work on modelling was done during 1984, and by the end of that year most product analysis had been completed. The computer projections were analysed and discussed with the IWP during 1985. In January 1986 the model structure and coefficients were adjusted, and all data updated to accept the latest data, in this case for the calendar year 1984.

Improved quality of this *Outlook Study*

FAO has carried out projections for the pulp and paper industry for many years, dating back to 1954. There is naturally a continuing effort at FAO to improve the quality of repeated studies of this type by taking full advantage of improved data, FAO's own increasing experience and expertise as well as the latest advances made in econometrics and electronic data handling.

This *FAO Outlook Study* benefits in many ways from the factors cited, and thus its value to the world's paper industry and the national authorities involved should be considerably enhanced.

Four key areas of improvement may be identified:

1. *Data*: FAO can now command uniform international data sets back to 1961. The longer data series naturally improves the performance of projections of this type.

2. *Control of data*: The recent advances in electronics capabilities have made a vast improvement in FAO's ability to control data and use them correctly. The sensitivity of the econometrics is thus considerably greater. This has allowed introducing important features not previously feasible, for example, in new models for production and supply.

3. *Defining uncertainties*: A major effort was made to clearly define the uncertainty behind the projections and the models. This has permitted better analysis and, in turn, presentation of such uncertainties.

4. *Conceptual framework*: The IWP was involved in the development of the original conceptual framework of this study. This improved the framework and the quality of the assumptions used for projections.

The nature of the projections

Central to the *FAO Outlook Study* are projections of the supply and demand of paper and the subsequent requirements for raw material inputs. "The term projections has been deliberately chosen to make it clear that these figures are not predictions but simply logical conclusions resulting from known facts about the past and explicitly stated assumptions regarding future development." This quotation from Egon Glesinger, former Director of Forestry in FAO and pioneer of FAO work on the outlook for the sector is a guideline to how these reference projections should be viewed.

General projections which cover production and consumption of all paper worldwide cannot be the sole basis for investment decisions. The data, because of its wide spread, can only start with input data nearly two years old. However good the assumptions, their accuracy will tend to fall off with time because no allowance can be made for political changes, exchange rate fluctuations, trade cycles or sociological attitudes.

Uncertainties over the statistical projections are compounded by those over the GDP forecasts used for the economic scenarios. The model gives better representation for regions and periods of years than for an individual country's consumption for one year. Moreover, the lack of sound worldwide historical data for the various subsectors of *other paper and paperboard* makes estimates in these areas less reliable and more dependent on the views of the IWP.

The detail consistently covered by available international data on the production and consumption of pulp and paper and on the performance of national economies sets limits on the refinement of relationships that may be identified through logical analysis. These calculated relationships are objective and unbiased by opinion or particular interest. The projections of the future derived from them are, statistically speaking, the most expected. However, as the relationships on which they are based are not precise, the projections must be recognised as subject to uncertainty, the degree of which is indicated by statistical confidence limits. There is also uncertainty related to the underlying economic forecasts.

Background documentation

Full detail of the projections are published in *Forest Products World Outlook Projections*, FAO, 1986. A selection of background documentation to the preparation of the FAO Outlook Study is published in *Readings from the Outlook Study for Supply and Demand of Pulp and Paper*, FAO, 1986. Annex IV lists the reports prepared during the study.

CHAPTER 2

RESULTS AND CONCLUSIONS

Paper is going to remain a growth industry. By 1995 total worldwide demand for paper and paperboard is projected to rise to between 246 million metric tonnes in the Chase scenario and 255 million metric tonnes in the FAO scenario. This represents an annual growth rate (from 1984 levels) of between 2.6 percent and 2.9 percent, and an increase in tonnage consumed annually of between 59 and 68 million tonnes from the 1984 level of 187 million tonnes (FIGURE 2).

The world average per caput consumption of paper and paperboard is projected to rise to 44 kg in 1995. This compares with 25 kg per caput in 1960 and 38 in 1984 (FIGURE 3).

The world's potential production capacity for paper and paperboard, projected to 1995, will be sufficient to meet projected demand at that time.

Supplies of major raw materials such as pulpwood, pulp, waste paper, chemicals, fillers and coating pigments, will be sufficient to meet the demand.

The *Outlook Study* projections for pulp indicate an increase in consumption from the 140 million ton level of 1984 to 177 to 181 million tonnes by 1995. However, the ratio of pulp content per ton of paper will decrease noticeably from 0.74 to 0.72. Waste paper consumption will increase sharply from 49 to 83 to 85 million tonnes, and will account for 32 percent of the total worldwide fibre supply to the industry in 1995.

Figure 3

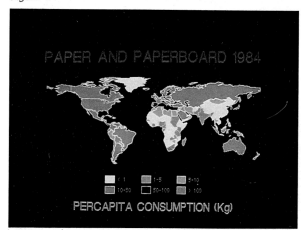

These conclusions about the growth of consumption and production are supported by an exploration of the three major product groups — *newsprint*, *printing and writing paper*, and *other paper and board* — and a review of greater depth of component products. The review covers developments in end-use areas, communications, packaging and transport.

The total volume of world trade in paper and paperboard is 39 million tonnes in 1984, projected to increase by 1995 to 55 million tonnes. If, however, total trade volume continues to increase at a rate slightly faster than total consumption — as it has recently — then a level of 60 million tonnes would be reached by 1995.

The volume of trade will increase. The detailed distribution of net trade flows depends on the relative expansion of production in different regions. The volume of net exports from developed countries may increase only slightly, while the dependence of developing countries on net imports may diminish by 1995.

Details of the major paper grades

The breakdown of major grades and the world-wide consumption of *total paper and paperboard* are illustrated in FIGURES 4 and 5.

Figure 2

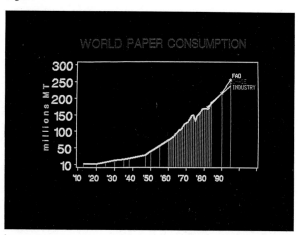

Figure 4

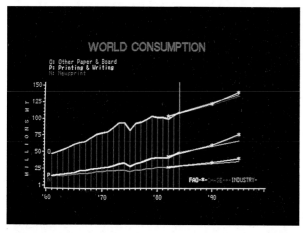

Figure 5

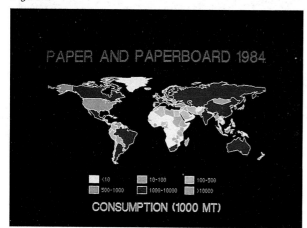

Newsprint

Developments in electronic data expected in the developed countries during the next decade may affect the growth rates of newspapers, as may more liberalised radio and television advertising policies in certain regions.

While in developed countries *newsprint* is used mainly for newspapers, in developing countries it has a much wider use, including school exercise and work books, business forms and even computer forms. Reduced basis weight in developed countries has tended to reduce *newsprint* tonnage consumption. Projected consumption of *newsprint* to 1995 is indicated at 39.3 to 39.7 million tonnes, representing annual growth rates from 2.8 to 3.1 percent. World consumption was 29.2 million tonnes in 1984.

The Industry Working Party (IWP) felt that of 27 developed and developing countries surveyed, 21 would have less consumption of *newsprint* by 1995 than indicated by either scenario. The IWP forecasts world total consumption at 36.4 million tonnes in 1995.

Printing and writing paper

Projected consumption of this sector is seen at between 73.6 and 75.5 million tonnes, according to the *CHASE* and *FAO scenarios*, respectively (1984 consumption was 48.8 million tonnes). This represents average annual growth rates of from 4.2 to 4.7 percent.

The IWP, however, forecasts total consumption to 1995 of only 67.7 million tonnes. The IWP believes that rapid growth in special-interest magazines is coming to an end; television advertising is expanding; and book and telephone directories show no growth. Electronic communications will take over new areas previously provided for by the print media.

There will be greater use of office papers. Print advertising will still increase, but higher postal charges will tend to favour reduced basis weights.

Information is a growth area, and despite the increase of electronic systems, the use of *printing and writing paper* will still increase, particularly in many developing countries, where there is a drive for greater literacy. A 1 percent improvement in literacy could result in a 1 percent increase in annual consumption of these grades in developing countries.

The *FAO Outlook Study* breaks down *printing and writing papers* into four main categories:

— Coated wood-free
— Coated wood-containing
— Uncoated wood-free
— Uncoated wood-containing

Coated wood-free papers now account for about one-third of the sector. Coated paper (wood-free and wood-containing) currently accounts for some 30 percent of the sector, and this share is projected to rise to 40 percent by 1995. This means generally higher average growth rates for coated papers (+5.6 percent) than for uncoated (+3 percent), over the period 1981-95.

Growth in *developing market economies* and *centrally planned economies* is generally projected to be higher than in the *developed market economies*.

Other paper and paperboard

With 109.2 million tonnes, this sector accounted for 59 percent of *total paper and paperboard* consumed in the world in 1984. The five components of this sector and their percentage of the sector in 1984 are:

	Percent
— Containerboard	40
— Folding boxboard	14
— Wrapping paper	18
— Household and sanitary papers	9
— Miscellaneous papers	19

Overall growth of this sector is projected to be 2.1 to 2.8 percent to 1995, to a total tonnage ranging between 132.9 and 139.4 million tonnes.

The growth rates in *developing market economies* and the *centrally planned economies* will probably be greater than those of the *developed market economies*.

While the breakdown of components of this sector is very important, the lack of data has meant that aggregated projections could only be prepared for developed countries.

Standard of living will be the main influence on packaging and wrapping grades, as well as the development of competing products (e.g. plastics). Developments in technology (bulk packaging reducing the need for paper) and the relative development of the many end-uses are other key factors.

The developed world is in the packaging age, where rationalization of distribution provides increased outlets for corrugated containers as well as shrink and stretch film.

In some developing countries paper containers are replacing wooden boxes and cotton sacks, but it is also commonplace to find that all sorts of paper and board may be used for wrapping, storage and other uses, and then re-used again. However, the developing economies have rapidly adopted the plastic bag and metal beverage can and, as income rises, may go through other phases of the packaging revolution that developed countries have known.

A look at the components

Containerboard: Linerboard and fluting of all types used in containers are included in this subdivision. The average growth for 22 selected developed countries is 3 percent. End-users are trying to reduce their overall packaging costs; distribution systems are being streamlined and returnable packaging being encouraged in some countries.

Folding boxboard: Solid bleached board; other folding boxboard and liquid packaging board are contained in this subdivision where the IWP forecast corresponds with the *CHASE scenario* projection.

Wrapping paper: Wrapping and packaging papers using sack kraft, vegetable parchment, sulphite wrapping and straw pulp are included here. Growth will be greater in developing countries, where wrapping papers are about 80 percent of the total of *other paper and paperboard*.

Household tissues and sanitary papers: Southern Europe will have a higher growth rate than Northwest Europe, where, like the United States, the market is showing signs of maturity.

Production trends for paper and paperboard

World production of *paper and paperboard* is projected to increase from 187 million tonnes in 1984 to 255 million tonnes under the *FAO scenario* or 246 million tonnes under the *CHASE scenario*, an increase in the range of 59 to 68 million tonnes (FIGURE 6). Some 40-46 million tonnes is the projected increase in *developed market economies*, 6-7 million tonnes in *developing market economies* and 11-14 million tonnes in *centrally planned economies*. The projected growth of production for *developed market economies* is also slightly above their projected growth for consumption, whereas for the other regions it is below consumption growth.

The future distribution of production is subject to uncertainty. The developed world may continue significant net exports to the developing world but the developing world may also continue to increase production faster than consumption, diminishing its dependence on imports.

At least to 1990 the projected growth of production is within the growth of capacity predicted by the industry in *FAO Pulp and Paper Capacities*.

Raw materials

Pulp consumption is projected to rise from 140 million to 177 to 181 million tonnes by 1995 although the share of pulp in the total furnish will decline. The average pulp input per ton of paper worldwide will drop from the 1984 level of 740 kg to 720 kg by 1995.

Waste paper use will increase substantially. Total input for paper production will increase from 49 to 84 million tonnes, and the share of waste paper in the world's total papermaking furnish will rise to 340 kg per ton of paper in 1995 from the 1984 level of 260 kg.

Figure 6

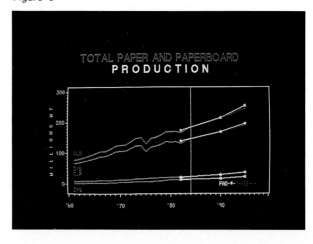

The IWP notes that increased use of waste paper may be more limited because additives, special coatings and laminates used in paper and paperboard converting, make an increasing proportion of waste paper unsuitable for recycling.

The IWP sees a small increase in pulp input to 760 kg with an expansion for thermomechanical and chemi-thermomechanical pulps, suggesting that they will replace standard mechanical pulps and chemical pulps. Other fibre pulps will continue to be important furnish components in developing countries.

On the basis of production, projections have been made of consumption of pulp and waste paper for paper production and the consumption of wood in pulp production. In developing countries, wood consumption is projected to increase by 12 million m^3 to 45 million m^3 but could almost double to around 60 million m^3. In developed countries the increase is about 20 percent to 1995, involving an increase in wood volume by 100 to 120 million m^3.

World trade in pulp and paper will grow

Total world trade in pulp and paper in 1984 was valued at around US$30 thousand million, and represented 1.5 percent of the world's total merchandise export. The tonnage traded included 39 million tonnes of paper and board, 21 million tonnes of mechanical and chemical pulps, and 4 million tonnes of waste paper. Most of this trade was in the developed countries.

In the developing countries, the net value of pulp and paper imports was US$4 thousand million, or 1 percent of their total imports.

Considering only *paper and paperboard*, in the 30 years to 1984, the total volume of trade increased from about 9 million tonnes valued at US$1.5 thousand million, to 39 million tonnes valued at US$20 thousand million. The growth in volume traded was, however, somewhat greater than the growth in consumption during this period, so that in 1984 some 21 percent of *total paper and paperboard* consumed was involved in international trade. The figure in 1961 was only 17 percent.

Future development of world trade in *paper and board* is expected to continue trends found in past periods. If the volume of *paper and paperboard* traded maintains the same proportions to total consumption as in 1984, namely 21 percent, then the total volume worldwide would increase to about 55 million tonnes by 1995. There is, however, a tendency for the proportion of trade in total consumption to increase slightly. If this trend continues, the volume of trade could be 60 million tonnes (FIGURE 7).

Figure 7

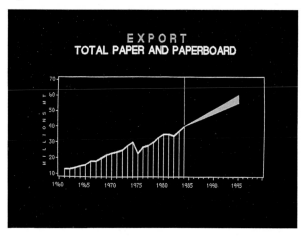

Trends in self-sufficiency

The world's pulp and paper market is made up of about 150 significant importers, 75 of which have some production and 45 of which are also exporters. This means 75 consuming countries are heavily dependent on imports for their paper supplies, and 70 of these are developing countries. These countries draw their supplies from the 44 exporting countries, though the main volume of supply is highly concentrated — five major exporting countries (> 1 000 000 tonnes/year) account for 66 percent of total volume traded (FIGURES 8 to 11).

In 1984, imports contributed 25 percent of developing country paper consumption. Imports of pulp and waste paper contributed some 15 percent of the fibre of their domestic paper manufacture.

Developing countries have grown in importance as importers of both paper and pulp. They have also entered the field as significant exporters, particularly of pulp. A feature of their trade in paper is the

Figure 8

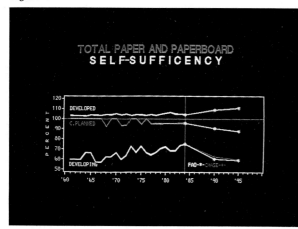

Figure 9

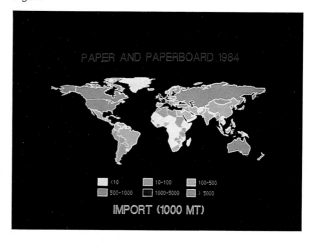

Figure 11

Figure 10

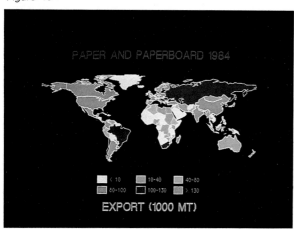

sumption. This would suggest that the net exports of the developed world would increase in volume, but by less than the 12-15 million tonnes indicated by the projections. Determining participation in the total volume of trade is the comparative advantage of the industry. Recent trends suggest that developing countries must be expected to participate to an increasing extent in the international trade in pulp and paper.

Product mix has changed

The product mix in world trade has changed dramatically. In 1954, two-thirds of the volume traded was *newsprint*, while by 1984 *newsprint* was only one-third of the total. Developing countries imported 1.4 million tonnes and exported 50 000 tonnes in 1954; by 1984 imports of these countries were 6.6 million tonnes and exports 1 000 000 tonnes.

Imports constituted about 60 percent of developing countries' total paper consumption in 1954. By 1984, only 25 percent was imported of a total consumption of 27.5 million tonnes. The exports of these countries were 5 percent of production in 1984. Five developing countries export more than 100 000 tonnes of paper: Brazil, China, Chile, Republic of Korea and Colombia.

predominance of other developing countries among their trading partners. Both paper and pulp are fields in which developing countries may be expected to further expand their participation, particularly as the volume of the domestic production expands. Although a large number of countries with small total consumption will remain heavily dependent on imports, the greater part of the need may be met from other developing countries.

In Chapter 6 alternative patterns in the relative development of regional consumption and production are considered. First those in which developed countries increase their level of net exports by 1995, and alternatively, following recent trends, developing countries increase their self-sufficiency. Developed countries' net exports would be in the range 18-21 million tonnes in the first case, compared to 9 million tonnes in the second.

Past tendencies suggest that the developing countries will increase their self-sufficiency through a more rapid expansion of production than con-

Trade in the different grade sectors

Although this *FAO Outlook Study* does not project trade volumes for individual grades or grade sectors, the history of trade development is reviewed here as a reference and guide to interpreting the projections of total trade volume. TABLE 1 summarizes world trade flows in 1984.

Table 1 - DIRECTION OF TRADE - PAPER AND PAPERBOARD (quantity 1 000 million tonnes)

1 9 8 4

MAJOR EXPORTERS across the top (EXPORTS); MAJOR IMPORTERS down the side.

MAJOR IMPORTERS	WORLD[1]	DEVELOPED	Finland	Sweden	USA	Germany, Fed. Rep.	Canada	France	Austria	Netherlands	Japan	USSR	Italy	Norway	Belgium-Lux	Others	DEVELOPING	Brazil	Others	WORLD IMPORTS[2]
WORLD	39 788	38 405	6 062	5 160	3 488	2 803	10 086	1 308	1 080	1 088	669	986	762	1 214	370	2 974	1 383	700	683	39 451
DEVELOPED	32 955	32 314	5 446	4 484	2 079	2 691	9 053	1 185	883	1 040	50	710	706	1 010	370	2 607	641	363	278	32 833
Germany, Fed. Rep.	3 697	3 640	755	929	110	—	100	458	324	343	3	30	224	193	166	5	57	57	—	4 429
United Kingdom	4 281	4 255	1 393	1 011	294	307	471	176	84	167	2	—	51	254	45	—	26	26	—	4 522
France	2 148	2 145	458	521	70	679	3	—	51	170	1	—	128	64	20	—	3	3	—	2 562
USA	9 425	9 302	441	232	·	168	8 115	26	—	42	16	6	120	122	111	—	123	123	—	9 849
Netherlands	1 542	1 521	279	348	93	430	16	96	48	—	1	—	19	74	28	—	21	21	—	1 688
Italy	1 134	1 069	100	233	129	244	17	159	112	18	—	—	—	29	—	—	65	65	—	1 405
Belgium-Lux	1 226	1 202	134	167	92	299	67	172	—	203	—	—	34	34	—	—	24	24	—	1 070
USSR	625	625	519	27	—	34	—	7	—	—	—	—	1	37	—	—	—	—	—	935
Japan	816	814	134	33	488	3	145	—	—	—	—	—	—	11	—	—	2	2	—	815
Denmark	686	686	247	310	1	57	—	4	—	23	—	—	—	44	—	—	—	—	—	743
Canada	591	589	39	18	500	17	—	5	—	3	—	—	6	1	—	—	2	2	—	538
Australia	623	599	179	46	107	36	86	6	—	5	14	—	53	6	—	—	24	24	—	634
Switzerland	319	319	63	106	17	86	—	22	3	10	—	—	4	11	—	—	—	—	—	418
Spain	217	217	86	39	—	43	—	19	—	7	—	—	13	7	—	—	—	—	—	241
German Dem. Rep.	196	196	4	2	—	20	—	—	—	—	—	170	—	—	—	—	—	—	—	277
Hungary	177	177	16	6	2	18	—	—	26	1	—	101	9	1	—	—	—	—	—	218
Austria	224	224	48	38	—	111	—	7	—	3	—	—	14	—	—	—	—	—	—	287
Others	5 128	4 734	551	418	176	139	33	28	235	45	13	403	30	122	—	2 541	294	16	278	2 202
DEVELOPING	6 833	6 091	616	676	1 409	112	1 033	123	197	48	619	276	56	204	—	722	742	337	405	6 618
Hong Kong	352	325	11	11	90	10	12	—	3	9	171	—	7	1	—	—	27	27	—	639
China	547	547	26	17	116	—	129	—	—	—	259	—	—	—	—	—	—	—	—	537
India	181	181	4	1	—	1	79	—	—	—	—	66	—	4	—	—	—	—	—	195
Singapore	164	155	15	30	42	7	—	—	—	9	48	—	—	4	—	—	9	9	—	268
Latin America	1 355	1 215	118	43	670	11	243	15	63	2	12	100	4	9	—	—	140	100	40	1 643
Africa	711	598	79	161	107	28	14	74	131	9	129	12	22	17	—	—	113	113	—	421
Others	3 523	3 070	363	413	384	55	556	34	—	19	—	98	23	169	—	536	453	88	295	2 915

[1] Volume identified with specific exporters.
[2] Volume reported by importers.

Newsprint. World trade in *newsprint* is dominated by the exports of Canada to the United States. Canada exported 8.1 million tonnes in 1984, and the Nordic countries exported 3.3 million tonnes. This combined total of 11.4 million tonnes accounted for 85 percent of the trade.

Some 43 countries produce *newsprint*, 15 of which are developing countries. However, half the world consumption is met from trade. This is due to the high concentration of production in Canada and the Nordic countries, accounting for 45 percent of the total, which in turn is largely exported to other countries. Developing countries, with imports of 2 000 000 tonnes in 1984, depend on imports for 40 percent of their consumption.

Printing and writing papers. World trade in 1984 was nearly 10 million tonnes. Developing countries' consumption in 1984 was 6 million tonnes, imports 1 000 000 tonnes and exports 400 000 tonnes. The net dependence on imports was 10 percent. In 1960, in a consumption of 1.6 million tonnes, dependence on imports was 20 percent.

Other paper and paperboard. World trade in *other paper and paperboard* in 1984 was 16.6 million tonnes, compared with world consumption of 109 million tonnes. Consumption in developing countries was 16 million tonnes, of which 2.5 million tonnes was dependent on imports, equivalent to 16 percent of consumption. For reference, in 1960 dependence on imports was 17 percent.

Trade in market pulp and waste paper

World trade in pulp (TABLE 2) in 1984 was 21 million tonnes valued at US$9 000 million. (This compared with about 6 million tonnes at just under US$1 000 million in 1954). Developing countries' imports were 2.8 million tonnes in 1984, while exports were 1.7 million tonnes. Net imports of pulp constituted 6 percent of developing countries' apparent consumption of 15 million tonnes of paper-grade pulp. This compared with 60 percent of a consumption of 700 000 tonnes in 1953.

The main developing country importers in 1984 were China (600 000 tonnes), Republic of Korea (400 000 tonnes), and Mexico, Argentina, Venezuela, India, Indonesia and Thailand, each with imports in the 100 000 to 200 000-ton range.

The main developed country pulp exporters are Canada (7.0 million tonnes), USA (3.2), Sweden (3.3), Finland (1.5), USSR (0.9), Portugal (0.6), Norway (0.5), New Zealand (0.4). Among major developing country exporters are Brazil (1.0 million tonnes), Chile (0.5), Swaziland (0.15).

Waste Paper Data (FAO, 1984), which assembled series from 1974 up to 1982, indicates total waste paper use in furnish that year of some 46 million tonnes. Of this, 5.3 million tonnes were traded. The net imports of developing countries were 1.5 million tonnes, about 7 percent of total fibre furnish for these countries and 25 percent of waste paper they utilized.

The largest exporter of waste paper was the USA exporting 2 million tonnes in 1982, followed by Federal Republic of Germany (0.6 million), France (0.4), Netherlands and Belgium (0.3); then Switzerland, Czechoslovakia, UK, Sweden and USSR, each exporting more than 100 000 tonnes. Significant importers in the developing world are China (500 000 tonnes), Republic of Korea (400 000 tonnes), Mexico (300 000 tonnes), Venezuela and Thailand (100 000 tonnes).

General observations on the industry

Sustained and economic fibre supply is an essential basis for efficient utilisation of the mill capital. Modern forestry technology is able to generate high yields of consistent quality, facilitating rapid return on the plantation investment and the concentration of the supply area. This potential has to be exploited in meeting the requirements of the paper industry. The industry's wood needs can be met in a way fully consistent with the physical, social and economic environment of many developing countries.

The locations suitable for this capital intensive industry requiring a sustained flow of inputs, water and energy are not widely available. Essentials, besides the access to wood and water, are the transport linkages for delivery of equipment, raw materials and the despatch of products, and the communications network to suppliers and to the market and to technical information. The industry requires a staff with relatively high levels of technical expertise, and locations which provide for accommodation and full support of this personnel.

Technical information across a wide field of specialisations is an essential resource for the efficient operation of such a complex industry. In a large scale enterprise, a part of this may be provided internally, but in all cases there must be efficient communications to a rich source of information on industry technology and research. This industry within countries has to develop a degree of such support. If it is beyond its capability to develop the support internally, then it should seek such support in national training and research institutions, or through linkages with institutions in other countries. A development of an efficient access to such costly, but essential, support provides a fertile area for international cooperation.

The pulp and paper industry has demonstrated a robust capability to develop its basic technology to respond to engineering innovation and the technological developments in the end-uses of its pro-

11

Table 2 - DIRECTION OF TRADE - WOOD PULP (quantity 1 000 million tonnes)

1 9 8 4

MAJOR IMPORTERS	WORLD[1] EXPORTS	DEVELOPED	Canada	USA	Sweden	Finland	USSR	Portugal	Norway	New Zealand	South Africa	Spain	Austria	France	Others	DEVELOPING	Brazil	Chile	Others	WORLD[2] IMPORTS
WORLD	21 375	19 687	7 040	3 215	3 271	1 561	888	662	574	471	368	311	257	241	828	1 688	976	493	219	21 369
DEVELOPED	18 095	16 987	6 421	2 373	2 843	1 371	732	616	537	313	268	308	257	234	714	1 108	766	199	143	18 540
USA	3 979	3 830	3 605	—	125	31	—	—	7	—	60	2	—	—	—	149	138	11	—	4 044
Germany, Fed. Rep.	2 455	2 323	514	440	640	352	45	84	108	—	3	83	26	28	—	132	79	53	—	2 781
Japan	2 052	1 835	805	669	117	39	—	—	1	202	2	—	—	—	—	217	194	23	—	2 149
France	1 115	1 059	219	179	348	120	—	91	53	—	—	49	—	—	—	56	24	32	—	1 611
United Kingdom	1 496	1 468	395	235	301	221	27	134	102	—	4	77	102	118	—	28	7	21	—	1 737
Italy	1 532	1 499	330	222	393	156	—	59	57	—	1	31	—	3	—	33	11	22	—	1 695
Netherlands	797	790	202	111	267	55	—	69	58	—	—	24	—	3	—	7	7	—	—	601
Belgium-Lux	725	438	99	131	73	35	—	24	49	—	—	9	—	18	—	287	265	22	—	448
Switzerland	220	220	31	35	73	35	—	10	8	—	—	21	—	7	—	—	—	—	—	290
South Africa	156	138	29	75	—	1	—	33	—	—	—	—	—	—	—	18	18	—	—	289
Spain	263	256	27	—	79	16	121	63	15	—	—	—	—	56	—	7	—	7	—	307
Hungary	149	149	—	—	14	5	44	—	2	—	—	—	7	—	—	—	—	—	—	220
Austria	153	153	15	24	54	5	—	5	6	—	—	—	—	—	—	—	—	—	—	247
USSR	185	185	—	4	34	121	—	—	26	—	—	—	—	—	—	—	—	—	—	222
Australia	223	223	66	18	3	20	—	5	—	111	—	—	—	—	—	—	—	—	—	230
Yugoslavia	190	190	14	7	12	2	64	—	—	—	—	—	91	—	—	—	—	—	—	214
Others	2 405	2 231	70	223	310	157	431	39	45	—	198	12	31	1	714	174	23	8	143	1 455
DEVELOPING	3 280	2 700	619	842	428	190	156	46	37	158	100	3	—	7	114	580	210	294	76	2 829
China	642	469	185	124	67	53	—	24	—	11	5	—	—	—	—	173	53	120	—	753
Korea, Rep. of	421	364	77	186	53	19	—	4	3	22	—	—	—	—	—	57	30	27	—	571
Indonesia	164	151	55	59	22	15	—	—	—	—	—	—	—	—	—	13	4	9	—	282
Mexico	286	277	39	237	1	—	—	—	—	—	—	—	—	—	—	9	9	—	—	221
Venezuela	241	152	52	95	4	—	—	1	—	—	—	—	—	—	—	89	43	46	—	230
Others	1 526	1 287	211	141	281	103	156	17	34	125	95	3	—	7	114	239	71	92	76	772

[1] Volume identified with specific exporters.
[2] Volume reported by importers.

11

ducts. The future lies in continuous evolution, both of production technology and in the markets served. Success is dependent to a high degree on adequate investment in research. The efficiency of the industry depends on the recognition that the highest levels of management and technological expertise have to be applied to every stage of the industrial and trading process. Access to this expertise has to be secured so that communications, training and research, and the necessary linkages, are the essential part of the expanding investment to meet the future needs for paper.

The value of international co-operation

No study can ever be better than the information on which it is based. The success of this *FAO Outlook Study* is largely due to active international cooperation which has allowed FAO to assemble accurate and timely statistics from all over the world. As a beneficiary of international cooperation, this will, through its usefulness over the coming years, hopefully help promote international cooperation even further.

There are still many areas where good information and statistics are lacking. It is hoped that individuals, enterprises and national institutions or authorities will both see the value of good communications and will be encouraged to even further efforts to improve international cooperation and communications.

This *FAO Outlook Study* provides a view of the future for the pulp and paper industry around the world. But remember that view must be re-examined continuously as the future becomes the present, as events modify the conditions or assumptions critical to this vision. The study and its projections should not be a static, passive instrument, but rather a dynamic tool for continuing analysis and improvement of our vision of the future.

CHAPTER 3

THE DEMAND FOR PAPER AND PAPERBOARD

This chapter starts with a brief description of past consumption of paper and paperboard. A discussion of factors which may have impact on the development of paper consumption in communications precedes the section which presents projections and forecasts of consumption of *newsprint*, and *printing and writing paper*. Likewise a discussion of factors affecting consumption of paper and paperboard in packaging and transportation precedes the presentation of projections for *other paper and paperboard*. The chapter ends with a summary of the projections and forecasts for the consumption of *total paper and paperboard*.

Please refer to Chapter 7 (Modelling Data and Assumptions) for full presentation of the economic scenarios and methodology employed in developing the demand projections in this chapter.

Past consumption of paper

World consumption of paper increased from 14 million tonnes in 1913 to about 40 million tonnes in 1950, and more than quadrupled to 187 million tonnes by 1984 (TABLE 3). From 1913 to 1950, the average annual growth rate was 2.8 percent. In the decade 1950-60, growth averaged about 5.8 percent per year; in the decade 1960-70 it was 5.7 percent and in the decade 1970-80 3.0 percent. The rate of growth in world consumption of paper was slightly higher than GDP growth from 1960-70, and slightly lower than GDP growth in 1970-80. Growth in paper

Table 4 - AVERAGE GROWTH RATES OF CONSUMPTION AND GDP By type of economy (percent per year)

Period	Developed market		Developing market		Centrally planned		World total	
	Paper	GDP	Paper	GDP	Paper	GDP	Paper	GDP
1961-70	5.3	5.0	9.0	6.0	6.5	6.8	5.7	5.4
1971-80	2.3	3.0	6.7	5.3	4.4	6.2	3.0	3.8

consumption was somewhat higher than GDP growth in *developing market economies* and slightly below GDP growth in the *developed market* and *centrally planned economies* (TABLE 4).

World average per caput consumption of paper increased from 25 kg in 1960 to 39 kg in 1984 (TABLE 5). The range in consumption is large: the average for *developed market economies* is 20 times greater than the average for *developing market economies*, and the variation between individual countries is even greater. The level of consumption bears a close relationship to the economic level. The world average consumption of paper has remained fairly constant at 15.5 tonnes per million dollars of GDP since 1960. The average is higher in *developed market* and *centrally planned economies*, but has declined slightly over the two decades. The average in developing countries was one-third of the world average level in 1960, but had increased to about half by 1984 (TABLE 6).

The 1984 percentage distribution of world consumption is: *newsprint* — 15 percent; *printing and writing* — 26 percent; and *other paper and paperboard* — 59 percent (TABLE 7). This distribution also holds more or less for the individual economic regions. The world average growth rates of consumption for the period 1970-1982 were: *newsprint* — 2.1 percent; *printing and writing* — 4.4 percent; and *other paper and paperboard* — 2.6 percent (TABLE 8). Regionally, the growth rates were lower than the world average in the developed market economies and higher in the developing and centrally planned economies. In the developing economies the highest growth rate was for *other paper and paperboard*.

Table 3 - TOTAL CONSUMPTION OF PAPER AND PAPERBOARD 1960-84 By type of economy (million metric tonnes)

Year	Developed market	Developing market	Centrally planned	World total
1960	61	4	8	73
1970	102.5	9.5	14	126
1980	129	18	22	169
1984	144	20	23	187

Table 5 - PER CAPUT TOTAL PAPER AND PAPERBOARD CONSUMPTION, 1960-84 By type of economy (kilograms)

Year	Developed market	Developing market	Centrally planned	World total
1960	95	3	8	25
1980	163	8	15	38
1984	177	8	15	39

Table 6 - CONSUMPTION PER UNIT OF GDP (CONSTANT PRICES), 1960 AND 1980 By type of economy (metric tonnes/million 1980 US dollars)

Year	Developed market	Developing market	Centrally planned	World total
1960	17.2	5.7	20	15.5
1980	17.0	8.9	18.2	15.7

Table 7 - COMPOSITION OF PAPER CONSUMPTION BY MAJOR SECTOR, 1984 By type of economy (million metric tonnes)

Product	Developed market	Developing market	Centrally planned	World total
Newsprint	23	3	3	29
Printing and writing	39	5	5	49
Other paper and paper-board	82	12	15	109
Total paper and paper-board	143	20	23	184

Table 8 - CONSUMPTION GROWTH RATES, 1970-82* By type of economy (percent per year)

Product	Developed market	Developing market	Centrally planned	World total
Newsprint	1.5	4.6	4.4	2.1
Printing and writing	4.0	6.2	4.3	4.2
Other paper and paper-board	1.6	6.3	3.7	2.3

* 1982 = 1980-84 average

Newsprint and printing and writing paper

The two product groups *newsprint* and *printing and writing papers* are mainly used in communications. *Newsprint* is used for newspapers, *printing and writing papers* are used in magazines, catalogues, advertising material, books, office stationary, for computers and copying, and in education. The aim in this section is to identify factors in the development of communications which may lead to a different course in paper consumption growth than that found in past periods.

Growth of communications

Recent economic development has been characterised by the growth of the service sector and the increase in the proportion of its employment concerned with information. According to a recent study (Hurwitz, 1984), employment in information activities in the United States indicated by employees in offices had reached 40 percent of the total work force by 1980; in Japan the corresponding figure is about 30 percent. Parallel trends are reported from many other countries. Associated with the expansion of the information activity, there has been a very rapid expansion in the flow of information.

Growth in words made available through 17 public media in the United States was, at a rate of 8.4 percent per year in the period 1960-77, more than double the rates of growth of the economy. But words consumed, that is actually attended to, grew at 3.2 percent. "More and more material exists but limitations on time and energy are a controlling barrier to people's consumption of words." The average American was consuming 48 000 of the 11 million words available daily. Words available in Japan grew at 9.7 percent, about the same as GDP. The average Japanese was consuming 23 000 of the 1.5 million words available.

The percentage of all communications supplied — "words transmitted" or carried by electronic media increased in the United States from 92 to 98 percent in the period 1960-80, whereas the supply through print media fell from 8 to 2 percent. In the same period the consumption — defined as "words attended to" — received from electronic media increased from 60 to 75 percent, while the proportion of consumption from print media fell from 30 to 18 percent. A similar tendency was found for Japan.

In terms of supply of information, nearly all the growth was in electronic media. Print media became increasingly expensive per word delivered, while electronic media grew cheaper. Though the main media — radio and television — produced the largest flow, the most rapid growth recently was in "point-to-point" electronic media, principally data communication between computers and the like.

It is useful to note a general conclusion which refers to the information overload, a tendency to fragmentation of the audience, trends to the use of more diversified and point-to-point media, and a shift from print to electronic media. "The significance of the trend may be reflected in new styles of use of information media including interactive retrieval, long-distance communications and intelligent processing of records".

Perhaps the important message from this Massachusetts Institute of Technology study of information flows concerns the very rapid growth it identifies, and the dynamics of the changing contribution of the individual types of media.

Advertising

An important factor in the growth of communications media in developed market economies is advertising expenditure. In 1983, advertising expenditure in the United States was US$75 thousand million, or 2.3 percent of gross national product (Coen, 1984). This constitutes an important contribution to financing the media. For newspapers advertising may account for 60 percent of total revenue, with only 40 percent coming from the sale of the newspapers.

Table 9 - DISTRIBUTION OF ADVERTISING EXPENDITURE IN THE UNITED STATES, 1950-84 (percent)

Media	1950	1960	1970	1980	1984
Television	3	14	18	21	23
Radio	11	6	7	7	7
Magazines	8	8	7	6	6
Newspapers	36	31	29	28	27
Direct mail	14	15	14	14	16
Miscellaneous	28	26	25	24	21

The distribution of advertising expenditure in the United States from 1950 to 1984 is indicated in TABLE 9. Since 1950 there has been a gradual reduction in the proportion of total advertising going to newspapers to the current level of just under 27 percent. In Europe, however, 48 percent still is allocated to newspapers (see TABLE 10). The percentage allocated to TV and radio is, understandably, greater in countries with more open access for advertising in those media.

In Europe as in North America there is evidence of redistribution of advertising to electronic media as restrictions are relaxed. There is evidence that specialised audiences tend to favour print media for advertising information. The overall growth of advertising is such that even with the continuation of such

a trend, the volume of advertising in print media would also increase.

Table 10 - DISTRIBUTION OF ADVERTISING EXPENDITURE IN EUROPEAN COUNTRIES, 1983 (percent of market)

Country	Newspapers and magazines	TV	Radio	Other*
Austria	50	30	13	7
Belgium	60	8	0.2	32
Denmark	57	0	0	43
Finland	67	9	0	27
France	59	16	9	16
Germany, Fed. Rep.	71	10	4	15
Greece	40	54	6	—
Ireland	47	32	13	5
Italy	44	30	5	21
Holland	59	5	1	35
Norway	71	0	0	29
Portugal	30	40	25	5
Spain	50	31	13	6
Sweden	51	0	0	49
Switzerland	64	7	0	29

* Excluding newspaper inserts, directories, films and direct advertising
Source: Euromonitor

Electronics and paper

Electronic media relate to paper as a vehicle for information in several ways besides simply providing competing media for entertainment and as an outlet for advertising (Oliver, 1984; Harris, 1984). Interactive systems such as videotex permit the selection of specific information by the consumer through electronic communications and visual display on a television screen or video. In France, for example, there is a project to replace the telephone directory by providing such an electronic system and installing the necessary video terminal with the telephone. This type of system would be in direct competition with printed directories, catalogues and classified advertisements. Conversely, the focus of interest generated by the electronic media, not to mention the need for guides to the use of the media, stimulates the publication of special interest magazines and books.

Growth in office use, both in industry and the service sector, generates more of consumption of stationery, forms, duplicating and copying paper, continuous stationery and printing papers. Two technologies have had particular importance in generating major increases in the flow of information and in paper consumption — in the office: photocopying/ duplicating, and the computer. Expansion, particularly in the use of computers, continues at a very rapid rate. Though advances in the technology may tend to reduce the amount of paper used by computers of a given capacity, there is a more than a compensating increase in the number and capacity of computers.

Literacy

Literacy is clearly an important factor in the consumption of *newsprint* and *printing and writing paper*. In *FAO's 1977 World Pulp and Paper Demand, Supply and Trade*, a quantitative assessment of a "literacy elasticity" was attempted, suggesting a positive value around unity — that is, a 1 percent improvement in literacy would be expected to result in a 1 percent increase in the annual consumption of paper. An average annual improvement of 1.6 percent in countries with a literacy level in the range of 25 to 75 percent was assumed in that study. Particularly, where in certain developing countries there is a policy to concentrate resources on education and to drive for an improvement in literacy, consumption of *printing and writing papers* will in turn be driven upwards. Income and population remain the underlying source for increase in consumption in developing countries. In some developed countries, the static or declining school population leads to a decrease in consumption in this area.

Industry Working Party views on media impact

In the foregoing section, a number of general features of the development of communications which affect the consumption of paper are presented. The Industry Working Party takes a pragmatic view of the direction and magnitude of the changes in paper consumption that might result. These judgements are set out qualitatively in the following paragraphs. Quantitative forecasts are compared with the projections in the sections on *newsprint* and *printing and writing paper* consumption.

Newspaper circulation

The lower growth in *newsprint* in later periods is expected because of the increasing tendency to shift from newsprint to other grades, and to some continuation of the movement to lower basis weight. Specific factors indicating a movement from newsprint include the reduction in newspaper circulation over the past 10 years in the United Kingdom, over the past 3-5 years in France and currently in the Federal Republic of Germany. There has, of course, been a resurgence in the United Kingdom with the recent introduction of new newspapers. As far as the United States is concerned, the decrease in circulation per household has, up to now, been offset by the growth in number of households. Over the last 3 years circulation per household has stabilised. Household formation will decline over the next 10-12

years, with the possible result that total circulation may turn down. However, the rate of household formation may be affected by increases in the number of single person households.

A number of perceptions of the possible development of newsprint consumption in newspapers emerge. First, it is considered reasonable to expect that at high levels of wealth and high relative levels of consumption, the marginal propensity to increase consumption will diminish, implying a decrease in income elasticity as the level of income increases.

Newsprint consumption based on newspaper circulation is considered to be strongly related to population dynamics. In countries where the population is static, circulation may diminish. In countries with growing literacy rates and expanding population, one might also expect increasing newspaper circulation. Several developing countries mentioned special policies to promote newspaper distribution.

Printing technology

The shift toward more offset and more multicoloured printing will probably not affect tonnage more than marginally. It is likely to increase waste, but on the other hand new printing methods and modern control systems are expected to make more efficient use of paper and hence reduce the printing waste considerably. This technical change will lead to an upgrading of newsprint quality and more tailor-made varieties of paper grades will be used. The variety of paper grades used for newspapers will expand in coming years. In addition, non-traditional areas for newsprint consumption are growing rapidly (see Printing and Writing section).

Electronic media

Competition of electronic media may increase and work against *newsprint* and *printing and writing paper* consumption, with advertising moving to electronic media, and electronic information storage and retrieval taking over from paper. On the other hand, electronic media may in some areas already be saturated so far as advertising is concerned and there may be a tendency to return to print media.

In most countries, print media have lost market shares of total advertising expenditures to electronic media, particularly to television. This is attributed to the change in habits toward use of television among the household-forming generations. New electronic media will take over an advertising role, and it is expected that restrictions in television advertising will be relaxed. In the last couple of years, however, the trend has been stabilised, at least in the United States. The single most important factor explaining this new development is audience fragmentation, mainly a result of the sharply increased variety of new media. The already well established media with good market coverage, (i.e. large daily newspapers), have strengthened their position and will continue to do so in the future. The major competition can be expected to take place between the various electronic media and thus will not have any negative effects on newsprint consumption.

It has been suggested that in developing countries electronic media might allow print to be bypassed. However, the developing country respondents indicate that they do not consider electronic media to be in significant competition with print. There is a very strong force to expanding consumption driven by the growth of literacy and the campaigns to improve the distribution of newspapers and the availability of educational material.

Printing and writing paper end-uses

Important growth areas mentioned by developed country respondents were specialist magazines, newspaper inserts, direct mail and catalogues. There is increasing use of newsprint by non-newspaper products. This can be important in developing countries. However, in developed countries this is combined with the use of higher grades of newsprint and the substitution of *newsprint* by higher quality *printing and writing paper* in these cases.

In Europe, most of the growth in *printing and writing paper* is expected from coated papers, used primarily for advertising purposes, as well as certain office papers.

As a general trend, basis weights of *printing and writing paper* have been reduced in nearly all end-use sectors. Improvements in paper quality and the pressure of rising postal rates as well as the general savings achievable in paper costs, have all led to this downward trend in basis weights.

There are large areas of traditional demand which are experiencing below-average growth, particularly SC (supercalendered) rotogravure paper for mass-circulation magazines in Europe. This is still regarded as a growth area in Japan, however.

Paper's strength for the specialised advertising media and its widespread use in the office will secure the future growth of the printing and writing sector, although it is facing some major competitive forces from elsewhere.

Some segments have been in decline for many years, the most important of these is writing paper. One reason is the decline in the number of school children in Japan and the major European countries (scholastic writing papers); another is the growth of alternative methods of communication, private as well as business.

Office papers consist of many grades for a great variety of end-uses. While some show considerable growth, others are either stagnating or even in decline. The general requirement for information in the office is on the increase, but at the same time forms of communication are changing rapidly. Growth of electronics and use of computers in business communication is accelerating and will continue to change the demand for paper. This has led to a growth in the demand for continuous stationery, paper for word processors as well as photocopying, while other grades like duplicating and stencil paper have been replaced.

Experience has shown that more efficient means of communication have led to an increase in the use of paper for the office. While the impact of future electronic systems will reduce the relative share of information transferred on paper, the "paperless office" which was so much talked about a few years ago will not impact paper demand within the time frame of this forecast. The high value of information relative to the cost of papers used, the sizeable investment costs needed to eliminate paper and the social attitudes toward paper will all secure a strong future for the demand of paper in the office. On balance, the growth of office papers is expected to grow more than the average for *printing and writing paper*.

The United States is on the threshold of implementing new technologies broadly referred to as Electronic Data Interchange, which will allow business, governmental agencies and other organizations to conduct business electronically. This action could result in a significant loss in market for business forms paper by 1995. Some say the only papers that will be replaced by Electronic Data Interchange are the documents moving through the United States Postal Service. However, the IWP believes the use of electronic transmission in business transactions will result in a major reduction of hard copy used internally in support of these transactions. As a result, the reduction in the amount of paper used within a company relating to business transactions will have a far greater impact on overall paper usage than the copies now mailed.

The most important trend for publication papers is their growing dependence on advertising as opposed to their entertainment uses. Papers used in advertising sectors are growing faster than books for example. In many end-use sectors there is a trend away from mass-circulation, wide-interest publications toward more specialised and well defined products.

The comment from Japan is that the younger generation is growing up in a visual age with a wider range of tastes and hobbies and a tendency to shun books. Nevertheless, inexpensive books are still selling well, but visual magazines with graphic and colour printing are displacing text-oriented magazines. These trends have led to an increase in the demand for colour printing and a greater emphasis on higher quality papers. This will continue in the future.

In summary, while it is not possible to clearly identify the impact of technological change in communications, the movements of advertising to new media and the development of electronic media in general mean that somewhat slower growth for paper in the *developed market economies* is considered probable in the future. Estimates for future consumption prepared by the IWP for a number of countries, taking these factors into account, are summarized later in this chapter.

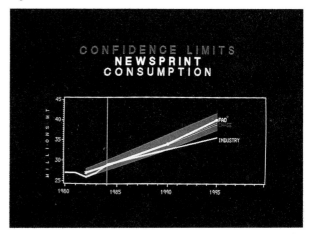

Figure 13

Newsprint consumption

World consumption of *newsprint* in 1984 was 29 million tonnes. (It was 14 million tonnes in 1961). Some 80 percent is consumed in *developed market economies*, with 10 percent in the *developing market*, and 10 percent in the *centrally planned economies* (FIGURE 12).

The projections of newsprint consumption (TABLES 11 and 12) indicate a recovery of growth rates above the 1970-1982 average of 2 percent to a world average of around 3 percent. World consumption in 1995 is projected at 39 million tonnes. The highest growth rates, approaching 5 percent by 1995, are projected for *developing market economies*, and would increase their consumption by 1.0 million tonnes to 5.4 million. Growth rates under the *CHASE scenario* are somewhat lower for all regions, particularly in the period 1990-95.

The statistical confidence limits suggest that, assuming the forecast of economic growth is correct, the projected world consumption in 1995 would be in the range 38 to 42 million tonnes (FIGURE 13).

Newsprint, mainly used for printing newspapers, is technically identified as containing 60 percent of

mechanical pulp and having a basis weight of 40 to 60 g/m^2. It has been separated in international statistics as a low cost paper whose availability stimulates the free communication of information.

Newsprint demand derives primarily from the household demand for newspapers, though some magazines, directories and manuals also use this type of paper. In many countries the distribution of these media is heavily dependent on their role in conveying advertising material. Household demand depends on the size and structure of the population, on various socio-economic and cultural factors, such as the level of literacy, and on the real income or purchasing power of the household. Support to distribution from advertising, and in turn the demand for distribution from advertisers, varies with the level of economic activity. The development of other media such as television and technological developments, both in communications in general and in paper-based communications also influence demand. Tonnage consumption has also been significantly affected in recent years by a reduction in basis weight of *newsprint*.

The price of *newsprint* has significance both in relation to the consumption level in individual countries and to the change in consumption over time. The declining real price in the 1960s may have contributed to the higher rate of consumption of growth in that period. Conversely the sharp rise from 1973, and the higher real price level prevailing in most years from 1973 to 1984, may have been a factor in the slower growth in that period.

Characteristic of the past 20 years' history of *newsprint* consumption have been very sharp fluctuations in consumption in many countries, particularly in the period 1975-80. Thus, when we consider the projections, the model provides a general indication of the most expected level and trend of consumption, but is subject to statistical error and also to the uncertainty associated with the economic forecasts used.

Figure 12

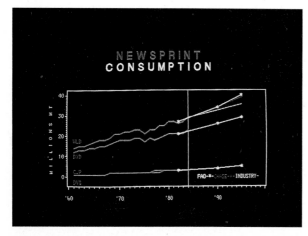

Table 11 - PROJECTED CONSUMPTION OF NEWSPRINT TO 1995 (million metric tonnes)

Type of Economy	Actual 1984	FAO scenario 1990	FAO scenario 1995	CHASE scenario 1990	CHASE scenario 1995
Developed market	22.9	25.7	29.3	25.8	28.9
Developing market	3.2	4.3	5.4	4.1	5.1
Centrally planned	3.0	4.0	5.0	4.2	5.1
World total	29.2	34.1	39.7	34.2	39.3

Table 12 - PROJECTED GROWTH RATES OF NEWSPRINT CONSUMPTION TO 1995 (percent per year)

Type of Economy	Actual 1970-82	FAO scenario 1982-90	FAO scenario 1990-95	CHASE scenario 1985-90	CHASE scenario 1990-95
Developed market	1.5	2.4	2.6	2.5	2.3
Developing market	4.6	4.3	4.8	3.9	4.5
Centrally planned	4.4	4.2	4.3	4.7	4.1
World total	2.1	2.8	3.1	2.9	2.8

Trend analysis

Demonstrating the use of trend analysis, the following growth trends to 1995 for newsprint consumption for major countries and regions (Graff, 1984) are compared in TABLE 13 with the *CHASE scenario* projections.

Table 13 - NEWSPRINT TREND GROWTH RATES (GRAFF) Compared with CHASE scenario rates (percent per year)

Region	1980-85 GRAFF	1985-90 GRAFF	1985-90 CHASE	1990-95 GRAFF	1990-95 CHASE
USA	0.7	0.3	2.4	-0.2	2.0
Western Europe	-0.6	1.7	2.0	-0.1	2.2
Asia	5.7	4.8	5.2	3.0	5.3
Latin America	1.8	1.2	2.5	0.5	3.8
Africa	-0.4	-2.0	3.0	-3.5	3.6
World total	1.1	0.9	2.9	0.1	2.8

The low trend growth rates indicated in this table result from the fact that the earlier part of the recent historical period 1960-70 had relatively high consumption growth, while the later years, 1970-80, tended to have lower growth. It should be noted that in spite of the recession years of 1982-83 world growth of newsprint consumption has so far averaged 2.5 percent in the 1980s.

Basis weight of newsprint

An important factor affecting the tonnage of newsprint consumption is the basis weight or grammage (TABLE 14 and FIGURE 14). After 1971, of newsprint basis weight fell from 52 g/m^2 to below 49 g/m^2 in the early 1980s, according to a study by the IWP. The fall in basis weight was steepest for the Nordic countries and Japan. It would be equivalent to a reduction newsprint tonnage consumption (at given demand for newsprint in m^2) of 6 percent for the USA and 8 percent for Western Europe, corresponding for the period 1970-80 to an annual decline of -0.6 percent for the USA and -0.8 percent for Europe.

Table 14 - SELECTED CHANGES IN AVERAGE BASIS WEIGHT OF NEWSPRINT AS REPORTED BY PRODUCERS, 1967-84 (g/m^2, year of change)

Country	1967	Main changes and year of occurrence			1984
Australia	52	49.7 (74)	48.8 (75)	48.1 (83)	47.9
Canada	51.9	51.6 (71)	49.3 (74)	48.5 (76)	48.4
China	51				51
Germany, Fed. Rep.*	52	50 (74)	49.5 (76)	48.4 (81)	48.3
Japan	51.8	50.4 (78)	48.7 (80)	47.5 (82)	46.6
New Zealand	52	48.7 (76)			48.6
Nordic Countries	52.1	49.2 (74)	47.1 (77)	45.4 (81)	44.8
United Kingdom*	52	51.0 (71)	48.0 (74)	47.3 (81)	47.3
United States	52.4	49.8 (74)	48.5 (78)		48.4
USSR	52	48.8 (82)			48.7

* Consumption

For the future, industry experts believe there will be variation in basis weight developments. For centrally planned countries, the main basis weight reduction is still to come. Some expect a slower trend toward decline in Europe and a higher one in the USA. Others feel that this decline will reach an eventual limit. Japan's projection indicates use of 46 g/m^2 will be 65 percent of consumption in 1986 and 80 percent in 1990. Both Japanese and American respondents point out, however, that this tendency may be limited as printability will have increased emphasis.

Industry Working Party forecast for newsprint

The IWP provided forecasts for newsprint consumption for 28 countries, and the total for 1995 for these countries was 27.0 million tonnes. This compares with projected consumption 30.3 million tonnes under the *FAO scenario* or 29.8 million tonnes under the *CHASE scenario* for the same 28 countries (TABLE 15).

Table 15 - NEWSPRINT CONSUMPTION: IWP FORECASTS COMPARED WITH PROJECTIONS TO 1995 (1 000 metric tonnes)

Country	Actual 1984	Projections FAO 1995	CHASE 1995	Forecast 1995
Canada	882	1 353	1 364	1 364
United States	11 912	14 252	14 578	14 000
Austria	220	223	212	185
Belgium	223	257	250	215
Denmark	173	217	210	160
Finland	203	234	229	240
France	578	827	766	550
Germany, Fed. Rep.	1 426	1 807	1 741	1 600
Greece	81	85	69	70
Iceland	4	6	6	7
Ireland	54	76	74	57
Italy	355	453	429	370
Netherlands	415	538	535	515
Norway	167	166	160	180
Portugal	35	53	49	52
Spain	256	292	291	240
Sweden	290	393	393	385
Switzerland	231	274	277	290
United Kingdom	1 456	1 756	1 847	1 350
Australia	568	761	762	700
Japan	2 783	4 509	4 019	3 000
Brazil	270	520	479	332
Chile	60	94	97	80/105
Colombia	78	150	131	124
Mexico	266	380	396	360
Indonesia	101	167	163	220
Philippines	85	162	110	106
Thailand	109	224	215	300
Total*	23 281	30 344	29 887	27 064

* 28 countries, not world total

Figure 14

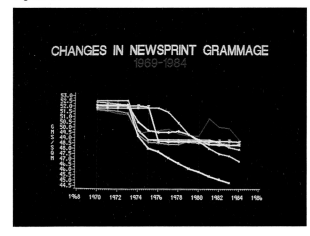

CHANGES IN NEWSPRINT GRAMMAGE
1969-1984

Printing and writing paper consumption

World consumption of *printing and writing paper* (other than newsprint) in 1984 was 49 million tonnes, 80 percent in *developed market economies* and 10 percent in each of the *developing market* and *centrally planned* country groups. *Printing and writing papers* are used in virtually all economic sectors for a wide range of publications and communications. Among the end-uses are: magazines, catalogues, directories, supplements, promotional material, books, continuous forms, photocopying, duplicating, educational writing, in-plant printing and general printing.

Printing and writing paper is a product area in which technological developments and the growth of electronic media developments have enormous implications, both by way of competition and in generating new uses and expanded demand. In recent years the more rapid expansion of communications than the general economy seems to have contributed to a high rate of growth of consumption of these types of paper.

Both the price level and the change in price over time may have contributed to the level of consumption. Similarly to newsprint, prices had a declining tendency in the 1960-75 period with a sharp increase 1974-75, then another tendency to decline.

Projections

The projections (TABLES 16 and 17) indicate similar growth rates in 1982-90 to the 4 percent average growth rate of the 1970s. World consumption in 1995 is projected to be 76 million tonnes under the *FAO scenario* and 74 million tonnes under the *CHASE scenario*. Growth in *developing market economies* and in the *centrally planned economies* is generally projected to be 5-6 percent, a higher

Figure 15

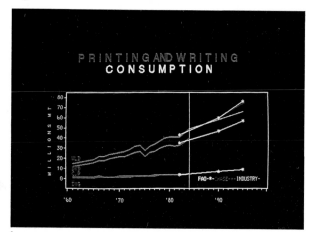

Figure 16

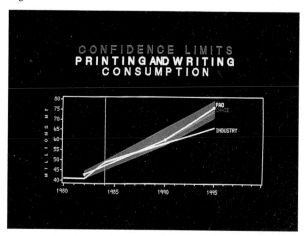

rate than in the *developed market economies* (FIGURE 15).

The *FAO scenario* growth rates exceed the 1970s rate in the period from 1990. The growth projected on the basis of the *CHASE scenario* is somewhat lower in all regions.

Projection for the world has statistical error limits of ±4 percent from 71 to 79 million tonnes. The confidence limits on projections for regions are wider, in the range ±6 to ±12 percent (FIGURE 16)

Table 16 - PROJECTED CONSUMPTION OF PRINTING AND WRITING PAPER TO 1995 (million metric tonnes)

Type of Economy	Actual 1984	FAO scenario 1990	FAO scenario 1995	CHASE scenario 1990	CHASE scenario 1995
Developed market	39.5	46.5	57.1	46.3	55.5
Developing market	4.8	6.8	9.3	6.6	8.8
Centrally planned	4.5	6.7	9.0	7.0	9.2
World total	48.8	60.1	75.5	59.9	73.6

Table 17 - PROJECTED GROWTH RATES OF CONSUMPTION TO 1995 (percent per year)

Type of Economy	Actual 1970-82	FAO scenario 1982-90	FAO scenario 1990-95	CHASE scenario 1982-90	CHASE scenario 1990-95
Developed market	4.0	3.8	4.2	3.7	3.7
Developing market	6.2	5.8	6.4	5.3	6.1
Centrally planned	4.3	5.8	5.9	6.2	5.6
World total	4.2	4.2	4.7	4.2	4.2

Disaggregated analysis

Printing and writing paper consumption may be broadly subdivided into four main categories determined by pulp input and pigment coating: coated wood-free, which accounts for 10 percent of total consumption; uncoated wood-free — 50 percent; coated wood-containing — 20 percent; and uncoated wood-containing — 20 percent of consumption. (Wood-free papers are made mainly of chemical pulps, whereas wood-containing papers include a proportion of mechanical pulps.) End-uses may be broadly grouped (see listing p. 22).

In view of the technical differences between these categories, it is of great interest to identify systematic variations in the development of consumption. Using detailed product data of the *OECD Pulp and Paper Statistics* it was possible to separate coated and uncoated printing and writing paper for *developed market economies*.

Coated paper accounted for one-third of the total consumption of *printing and writing paper* in these

Figure 17

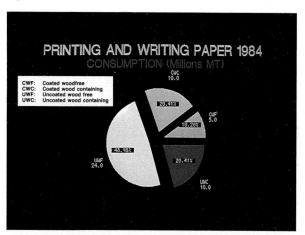

Paper Category	End Use
Coated wood-free	Special interest magazines, catalogues, books, inserts and promotional papers and company reports — for products requiring very high quality print
Uncoated wood-free	Photocopying, laser printing and duplicating, continuous stationery, books, educational writing, envelopes, and general printing
Coated wood-containing*	Magazines and supplements, direct mail and mail order and catalogues, inserts and other advertising material, books
Uncoated wood-containing	General interest magazines, newspaper supplements, directories, paperback books, catalogues, inserts, continuous stationery and duplicating

* Includes supercalender and lightweight coated papers.

Figure 18

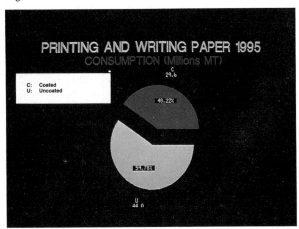

The analysis suggests a somewhat smaller differential in the growth of coated and uncoated in the case of North America and Japan compared with the total for *developed market economies*.

Industry Working Party Forecast: The IWP forecasts for a number of countries are shown in TABLE 19. An IWP Subgroup formulated forecasts for Europe by taking account of expected growth in particular end-uses, the competition from electronic media for advertising and the expected inroads of electronic communications on the use of paper in the office. The Japanease industry contributed a

Table 19 - CONSUMPTION OF PRINTING AND WRITING PAPER Industry Working Party forecasts compared with projections (1 000 metric tonnes)

Country	Actual 1984	Projections 1995		IWP Forecast 1995
		FAO	CHASE	
Canada	1 372	2 038	2 061	2 061
United States	18 085	22 869	23 268	23 125
Austria	100	202	188	105
Belgium	586	841	809	690
Denmark	224	347	333	305
Finland	431	647	628	360
France	2 422	3 993	3 601	2 960
Germany, Fed. Rep.	3 927	5 893	5 604	4 540
Ireland	34	60	58	55
Italy	1 782	2 470	2 486	1 920
Netherlands	743	1 048	1 041	840
Norway	153	284	271	215
Portugal	133	161	143	190
Spain	707	1 183	1 177	1 085
Sweden	681	966	966	750
Switzerland	386	570	580	450
United Kingdom	2 186	2 838	3 012	2 650
Australia	407	748	750	810-850
Japan	4 362	8 352	7 309	5 355
Brazil	787	1 755	1 617	1 342
Chile	74	102	106	71-85
Colombia	76	178	156	167
Mexico	559	937	846	726
Indonesia	192	391	382	446
Philippines	45	126	85	77
Thailand	46	146	140	188
Total*	40 500	59 322	57 617	51 510

* Selected countries only, not world total

countries in 1981. During the 1970s, growth of coated printing and writing paper has been much more rapid than that of uncoated in practically all *developed market economies*.

For the *developed market economies, printing and writing paper* was made up in 1981 of 32 percent coated and 68 percent uncoated paper. Using the disaggregated model, the detailed analysis suggests that the shares might change to 40 percent coated and 60 percent uncoated by 1995 (FIGURES 17 and 18). Applying this to the aggregate projection for *printing and writing paper* the following composition of projected consumption is obtained, as shown in TABLE 18.

Table 18 - PRINTING AND WRITING PAPER CONSUMPTION IN DEVELOPED MARKET ECONOMIES IN 1995

Product	Consumption (million metric tonnes)		Annual growth rate (percent)
	1981	1995	1981-95
Coated	10.3	22.0	5.6
Uncoated	22.1	33.2	3.0
Total	32.4	55.2	4.0

review of the expected growth of particular end-uses in that country. For the countries for which it made forecasts, the IWP expected consumption in 1995 of 51.5 million tonnes compared with 59.3 million tonnes under the *FAO scenario* and 57.6 million tonnes under the *CHASE scenario*.

Other paper and paperboard

World consumption of *other paper and paperboard* in 1984 was 109 million tonnes, of which 82 million tonnes was in the *developed market*, 12 million in *developing market* and 15 million tonnes in *centrally planned economies*. This broad category heading accounts for 60 percent of all paper and paperboard and embraces a very wide range of types and end-uses, including all wrapping and packaging paper plus containerboard, folding boxboard, sacks papers, household and sanitary papers and a wide range of speciality papers.

The composition of world consumption in this broad category is roughly as follows:

Main component sectors	Consumption 1984 (million tonnes)
Household and sanitary	10
Wrapping and packaging	78
Containerboard	(43)
Folding boxboard	(15)
Wrapping	(20)
Other (not specified above)	21
Total other paper and paperboard	109

Detailed international statistical information even on this breakdown is not available. Because of differences in production processes and raw materials associated with these products, and the diversity in the end-use areas where the products are consumed, developments in countries for which information is available have been explored. This brings out features which suggest the direction of future development in those countries and in other regions.

The contents of this section are:

1. Review of packaging developments.

2. IWP reviews of all grade sectors comprising the category.

3. Comparison of projections and forecasts.

Developments in packaging

Wrapping and packaging form the major part of the heterogeneous category of *other paper and paperboard*. In a way, this category has displayed the greatest deviation from earlier growth patterns. While *newsprint* and *other printing and writing paper* have broadly followed the path of general economic development, the growth rate of this category was sharply retarded in the late 1970s.

Packaging has the functions of protecting products, facilitating handling of products at all stages from production to delivery to the consumer, and of communicating information about the product and its maker. These functions can be performed in different ways, using different materials, and the selection of a particular material or combination of materials will depend on cost effectiveness in performing the function. The latter is a complex of cost of material, cost efficiency in the packaging process, efficiency in handling, space utilization, and weight. In all this is the interplay of costs of labour, storage and transport, not to mention the benefit perceived by the consumer.

A description of the recent history of packaging in Sweden (Bergstedt, 1985) illustrates some features of packaging development.

"The paperboard container age began in the mid-1960s to meet the requirements of the self-service shops for easy handling and good product display through printing and information. Another reason for the increase in paperboard was that deepfreezing techniques had been perfected. The paperboard milk package had cornered 100 percent of the milk market.

"The plastic and the beer can age followed in the late-1960s and early 1970s. The use of polyethylene shrink film for wrapping both pallet loads and transport packages for food, detergents and other consumer products increased rapidly, as this new technique was found to give low-cost convenience in distribution. Plastic bags and laminates were found to give better protection at a low cost. During this period, the use of metal cans for beer and fruit drinks was found to be a convenient product in comparison with the glass bottle, regardless of whether the latter was returnable or non-returnable. Although the metal can was much more expensive for the consumer to buy, its convenience for carrying and for keeping the product fresh and cold made it less sensitive with respect to price.

"In recent years, we have seen what could be called the distribution packaging age: an era characterized by new ways of rationalizing distribution of goods from producers via wholesalers to retailers. Here we can see new and increased outlets for corrugated containers and polyethylene shrink film and stretch films. This trend will surely continue."

It is noteworthy that from 1965 to 1980 paper and paperboard maintained a market share in Sweden in an increasing market (TABLE 20).

24

**Table 20 - CHANGES IN PACKAGING
CONSUMPTION IN SWEDEN, 1965-80**
(percent share of total Swedish market
expenditure)

Material	1965	1970	1975	1980
Paper and paperboard	45	47	48	47
Plastics	11	15	21	24
Glass	7	7	5	4
Metal	19	19	18	17
Wood and others	18	12	8	8
Total value *	1 425	2 125	3 300	4 650

* Million Swedish kronor (1982 prices)

There appear to be four general developments in packaging technology which will have significant impact on the use of paper and paperboard. The general search to reduce weight or bulk of packaging material may favour the replacement of wood or glass by paper, but this may be replaced in turn by plastic. Secondly, one may identify the movement toward integrated systems which require the compatibility of the packaging material with the packaging process, and further with the means of storage, transportation and display. Frequently this involves a combination of packaging and packaging materials. Thirdly, there is the movement toward cost minimisation, which may involve reduction in material utilised or the substitution of less costly material. Finally, there are environmental concerns favouring returnable and reusable containers, recycling of material or use of biodegradeable material (Jerkeman, 1984).

The stiffness requirement of a rigid packaging material is achieved in the cheapest way by using paper. This is born out by the predominant role of paperboard containers and the folding box. Efficiency of rigid containers may, however, be further improved by using of a combination of paperboard and polysterene. A great weakness of cellulose fibre, and in turn the corrugated paperboard container, is its vulnerability to moisture. Thus a further improvement in the effectiveness of the package is obtained by use of a protective wrapping of stretch plastic.

The use of such combinations of packaging materials are characteristic of current developments. Another area of development is the flexible container in which paper is combined with aluminium foil or plastic film to provide combinations of printability, moisture barrier, chemical resistance, gas-tight qualities and possibility of heat sealing. A very current development is the use of paperboard to provide a protective box around an inner packaging, meeting barrier requirements. This inner packaging may be paper, plastic, metal foil or a combination of these.

A development in packaging and transport is the extension of bulk transport and intermediate bulk containers which meet requirements for protection of sensitive commodities such as sugar and cement.

Once the technical requirements are met, the determining criteria relate to system cost. The paperboard container has held comparative advantage in meeting stiffness requirements in packaging. There is, however, direct competition between paper and plastics in various sack markets. An example is the competition between the plastic and the paper refuse sack. Development of a 100-g/m^2 paper sheet with the necessary high stretch and wet strength properties has put paper back into competition in this market, where the cost of plastic is critical.

Investment in research and development, which is at a relatively high level in the plastics industry, has an important role in determining the availability of competitive products and processes.

Industry Working Party reviews containerboard

Containerboard is the case-making material made of linerboard plus fluting or corrugating medium. The IWP and its Subgroup on Containerboard (Meister, 1984) singled out the following qualitative factors for particular importance in Containerboard development.

— Attempts by end-users to reduce overall packaging requirements to control packaging costs,
— Streamlining of distribution systems and inventory control procedures that reduce packaging needs,
— Government regulations that encourage the use of returnable packaging in some countries,
— Saturation of traditional markets for containerboard in some countries, counterbalanced by growth in newly "opened" markets in others, (e.g. fruits and vegetables) currently using other materials,
— Loss of some markets to competing packaging materials, particularly plastic film.

In Europe the traditional markets for corrugated board are either approaching or have reached saturation, and only a few "new" markets can be penetrated. The Subgroup points to the loss of some markets to competitive packaging materials, trends to lighter basis weights, and the general trend towards cost saving in the use of packaging materials. Government legislation encouraging the use of returnable packaging materials is also cited.

The United States forecast is influenced by the following considerations:

— The streamlining of distribution systems, like "just-in-time" delivery, as well as innovations in inventory management, which may reduce packaging requirements.

— Plastic films and bulk packaging will make some gains in market penetration over the forecast period, displacing containerboard.

— The growth in consumer goods production will outstrip demand for containerboard during the 1985 to 1995 period.

— Basis weight changes are likely to be minimal, at least over the forecast period.

— More efficient use of containerboard in box manufacturing will reduce the waste factor by 1 percent per year from 1990 to 1995, or a cumulative reduction of 5 percent.

The IWP forecast indicates somewhat more rapid growth for Japan and slightly lower growth for Europe, Canada and the United States than is shown in the disaggregated projections. They are compared in TABLE 21. The disaggregated projections are discussed later in the chapter.

Table 21 - CONTAINERBOARD CONSUMPTION, 1995
Industry Working Party forecasts compared with disaggregated CHASE projections (million metric tonnes)

Country	Actual	IWP Total	Disaggregated CHASE
	1981	1995	1995
United States	17.3	24-25	25.4
Canada	1.2	1.7	2.0
Japan	4.6	6.6	5.4
EEC	7.1	9.8	10.7
Scandinavia and other Western Europe countries	1.9	2.6	3.3
Total developed market economies	32.1	44.7-45.7	46.8
South America	1.3	3	—

Folding boxboard

Plastic materials have made inroads into the traditional uses of carton. This tendency could very well continue, especially if cost relations between raw materials change in favour of plastics as packaging based on a combination of board and plastic increases. The growing demand for packaging for convenience food and ovenboard (for microwave ovens) should give cartonboard new opportunities (Thornander, 1985).

The trend toward lower basis weights has already resulted in a situation where little further development can be expected. Too much of the material's rigidity would then be lost.

In Western Europe there seems to be a relatively balanced situation between the use of virgin-fibre and waste-based board over the last few years. The existing food packaging regulations in Western Europe are not expected to lead to any change in the usage of different types of board. There is, however, for reasons of hygiene, a trend to use virgin fibre or pre-consumer waste paper for some specific papers and boards, which come into long-term direct contact with wet or greasy food-stuffs. Another trend is toward the combination of materials such as board/aluminium foil or board/plastic film.

In the United States, the Food and Drug Administration mandates strict rules for the use of recycled board in food packaging, including rules for use when combined with a plastic barrier. A slight decline in the proportion of waste-based board is perceptible and less growth is expected here.

Wrapping and packaging grades

A Japanese long-term forecast (Japan Paper Association, 1984) comments: "Among packaging papers, the growth rate for unglazed kraft paper which accounted for 55 percent of the sector, started to decline in the early 1970s, and demand dropped substantially following the second oil crisis into the 1980s. The main reasons were: (1) overall decline in demand due to recession; (2) a general trend toward cost saving, weight saving, rationalization, simplification and the elimination of packing; (3) the promotion of bulk transportation and mass transportation technology to reduce distribution costs; and (4) competition from petrochemical packaging materials, especially the expanding use of high and medium density polyethylene bags in the square bottom paper bag sector. Factors which may lead to a demand increase include such as the trend toward sophisticated carrier bags and square bottom paper bags made of bleached kraft paper."

Several developing countries mentioned the replacement of wooden boxes with paper containers and the replacement of cotton sacks by multiwall paper sacks. In some countries paper sacks have been replaced in supermarket use by plastics, but multiwall sacks, boxes and cases continue to be favoured. The problem of plastic pollution was mentioned as a concern, and the disadvantage of the plastic sacks resulting from risk of puncture and difficulty of stacking lead to the multiwall paper sack being favoured. A significant feature of consumption in developing countries is the frequent re-use of *newsprint, printing and writing paper* and wrapping/packaging papers and board for wrapping, storage and other uses.

Household and sanitary papers

Household and sanitary papers include a range of different product categories, such as toilet paper, towels, industrial wipers, handkerchiefs, facial tissues, serviettes, kitchen wipers, baby napkins, other napkins and pads, and cellulose wadding for medical purposes. From a viewpoint of paper grades, *household and sanitary papers* include cellulose wadding, one side-glazed paper (hard tissue), creped paper and soft tissue paper. (Note: Products made from fluff pulp are not included, nor are non-woven fabrics.)

An IWP Subgroup (European Tissue Symposium, 1985) considers that the market in Europe shows signs of maturity, having reached maximum consumption, and is characterized by heavy substitution. On the one hand paper, substitutes for other materials, mostly textile, on the other hand paper is substituted for by other materials, mostly cellulose fluff and non-woven fabrics. In all Western European countries, for example, baby napkins made from paper have replaced textile napkins, while paper napkins, themselves, are being replaced by napkins consisting mainly of fluff and non-woven fabrics. As the substitution by fluff pulp and non-woven material started at a much lower level of consumption, Western Europe may not reach the per caput consumption level of *household and sanitary papers* of the United States, and low income countries may not experience the per caput consumption of the high-income countries. Aside from the type of substitution mentioned above, very little change in usage patterns is expected in the future.

The trend forecast proposed by the European Subgroup is summarised in TABLE 22 and compared with the disaggregated CHASE projection. National estimates are also provided for the United States, where the projected growth rate of 1 percent/

year is regarded as appropriate, for Japan, based on a growth rate of 2.8 percent, and for Australia. The disaggregated projection is discussed later in this chapter.

Projections and forecasts, and their comparison

The following sections consist of aggregate projections for *other paper and paperboard*, the disaggregated analysis, and a comparison of the IWP forecasts with the projections.

Projected consumption

World consumption of *other paper and paperboard* is projected to grow from 109.2 million tonnes in 1984 to 139.4 million tonnes in 1995 under the *FAO scenario* or 132.9 million tonnes under the *CHASE scenario*. The overall annual growth rate rises from 2 percent to nearly 3 percent under the *FAO scenario*; while it remains around 2 percent under the *CHASE scenario* (FIGURE 19).

Growth is projected to be lower than average in the developed market economies: 1.2 percent to a maximum of + 1.6 percent under the FAO scenario, while growth in developing market economies is projected at 5-6 percent under the *FAO*, and 4-5 percent under the *CHASE scenario* (TABLES 23 and 24).

Table 22 - HOUSEHOLD AND SANITARY PAPER CONSUMPTION Industry Working Party forecasts compared with disaggregated CHASE projections

	Actual	Trend growth rate (percent per year)			Forecasts (1 000 m tonnes/year) Disaggregated	
	1980	1980-85	1985-90	1990-95	1995	1995
Germany, Fed. Rep.	517	2.8	1.0	-0.8	598	912
France	234	6.7	5.2	3.7	500	386
United Kingdom	460	2.1	0.7	-0.7	511	832
Italy	350	5.5	4.0	2.4	625	415
Nordic countries*	278	2.9	0.9	-1.1	316	484
Other	535	6.8	4.0	1.1	952	1 385
Total Western Europe	2 375	4.4	2.7	0.8	3 503	4 414**
Japan	896				1 344***	1 487
Australia	106				160***	273
United States	3 865				4 830	4 830
Total	7 242				9 837	11 004

* Denmark, Finland, Norway, Sweden
** Average projected growth 1980-85 1.8 percent, 1985-90 5.2 percent, 1990-95 5.4 percent
*** National projection

Figure 19

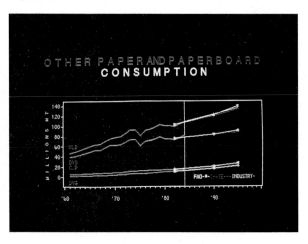

Figure 20

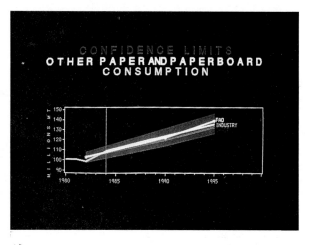

The statistical limits on the projection are ±5 percent for the world and in the range ±5 to ±13 percent for regions. These limits suggest that projected world consumption in 1995 would be within the range of 126 to 146 million tonnes (FIGURE 20).

Table 23 - PROJECTED CONSUMPTION OF OTHER PAPER AND PAPERBOARD TO 1995 (million metric tonnes)

Type of Ecomony	Actual 1984	FAO scenario 1990	FAO scenario 1995	CHASE scenario 1990	CHASE scenario 1995
Developed market	81.8	83.8	90.6	83.6	88.1
Developing market	12.0	16.7	22.2	15.7	20.6
Centrally planned	15.2	20.9	26.5	20.2	24.1
World total	109.2	121.6	139.4	119.7	132.9

Table 24 - PROJECTED GROWTH RATES OF CONSUMPTION OF OTHER PAPER AND PAPERBOARD TO 1995 (percent per year)

Type of Economy	Actual 1970-82*	FAO scenario 1982*-90	FAO scenario 1990-95	CHASE scenario 1982-90	CHASE scenario 1990-95
Developed market	1.6	1.2	1.6	1.1	1.0
Developing market	6.3	4.9	5.8	4.1	5.5
Centrally planned	3.7	4.4	4.8	3.9	3.6
World total	2.3	2.1	2.8	1.9	2.1

82* = 1980-1984 average

Disaggregated projections

As mentioned in Chapter 7, consumption models were prepared with detailed economic indicators: household and sanitary paper was related to aggregate GDP, folding boxboard to private consumption, and containerboard and wrapping paper were related to industrial production. Absence of data on prices and the short time series prevented consideration of either a relationship to price or the inclusion of a time trend in the analysis. Furthermore, this analysis could only be completed for *developed market economies*.

Analysis of the four component sectors of other paper and paperboard indicates that the fastest growing product would appear to be household and sanitary paper, with a projected average annual growth of 3-4 percent.

Of the wrapping and packaging papers, fastest growth is projected for containerboard with an average of 3 percent — slightly higher than average at 4 percent for Europe, and nearer to 1 percent for Japan. Similar growth is projected for folding boxboard for Europe and Japan, but a much lower rate for the USA. Growth of wrapping paper is projected at about 1 percent on average — with a very low rate for Japan (0.2 percent) and a rather high rate for the United States (2 percent).

The sum of the disaggregated projections of these four components results in a somewhat higher growth and total consumption by the year 1995 than the aggregate projections for *other paper and paperboard*. Under the *CHASE scenario*, the total of disaggregated projections for *developed market economies* is 100 million tonnes, while the aggregate projection is only 88 million tonnes.

In view of the previously mentioned limitations, it is perhaps best to regard the disaggregated analysis as a basis for estimating the development of shares of the component products (FIGURES 21 and 22). In TABLE 25 growth rates from the disaggregated

Figure 21

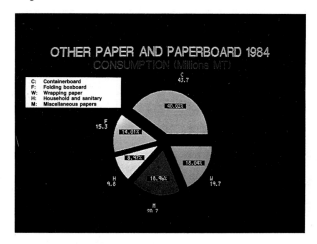

Figure 22

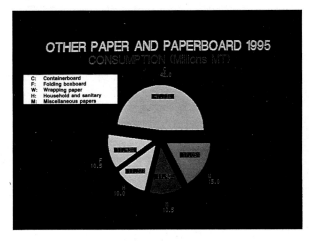

analysis are compared with the growth rates in the aggregate projection for *other paper and paperboard*. The share of the total of *other paper* from the disaggregated analysis is allocated to the total growth projected in the aggregate analysis. New growth rates for the component products are then calculated, which are consistent with the growth in the aggregate projection.

As noted in the discussion of containerboard, the *developed market economies* included in the IWP review are forecast to have a consumption in 1995 of 44.7 — 45.7 million tonnes. The disaggregated projection for those countries is 46.5 million tonnes, or 1-2 million tonnes more. The allocation (TABLE 25) the aggregate projection, suggests a total of 42 million tonnes, 6 million tonnes below the disaggregated projection and 4 million tonnes below the IWP forecast.

Household and sanitary paper consumption in 1995 is forecast, for the countries included, at 9.8 million tonnes, compared with the disaggregated projection of 11 million tonnes. The allocation in

TABLE 22 of the aggregate projection is 10 million tonnes, closely approximating the IWP forecast.

Summary forecast for other paper and paperboard

The IWP's review of components of *other paper and paperboard* is not complete either in product coverage or for countries. It was not possible to deal comprehensively with wrapping and sack papers. The heterogeneous collection of paper and board which are not packaging and not household — the "other not elsewhere specified" group was not tackled. However, Japan, Australia and several countries from Asia and Latin America provide forecasts for the total of *other paper and paperboard*, and the IWP Subgroups covered 80 percent of USA consumption and 45 percent of European consumption.

TABLE 26 compares the IWP forecasts with the projections of consumption of *other paper and*

Table 25 - OTHER PAPER AND PAPERBOARD CONSUMPTION TO 1995 Comparison of disaggregated and aggregate projections for developed market economies

Product	1981	Actual	Disaggregated projection			Aggregate CHASE projection	
			Growth	1995			Growth
	Total	Share	1981-95	Total	Share	Total 1995	1981-95
	mt	percent	percent	mt	percent	mt	percent
Household	8	11	2.6	11	11	10	1.4
Container	32	43	2.6	48	48	42	2.0
Folding boxboard	10	13	1.1	12	12	10.5	0.4
Wrapping	15	20	0.8	17	17	15	0.0
Other	11	14	0.7	12	12	10.5	0.3
Total	76	100	2.0	100	100	88	1.1

Table 26 - CONSUMPTION OF OTHER PAPER AND PAPERBOARD IN 1995 Industry Working Party forecasts compared with projections (1 000 metric tonnes)

Country	Actual 1984	Projections 1995		IWP Forecast 1995
		FAO	CHASE	
Canada	2 421	2 601	1 625	2 625
USA	38 731	39 499	40 717	41 300-42 500
Japan	12 276	16 246	14 136	14 500
Australia	1 121	1 214	1 217	1 500-1 600
Brazil	2 224	4 539	4 001	4 539
Chile	164	225	240	320-360
Colombia	293	569	465	480
Indonesia	327	548	529	890
Philippines	220	515	281	330
Thailand	324	691	650	740
Western Europe	25 693	28 905	27 732	30 100-35 700
Total	83 794	95 552	92 593	97 324-104 264

paperboard. For USA and West Europe, the IWP forecasts for *household and sanitary paper, containerboard and folding boxboard* are augmented (a) by adding the rest of other paper, unchanged from the 1984 amount and (b) by adding the rest, increased at the same rate as the products included in the forecast. It is important to note that in the resulting comparison for these products, the IWP forecast generally relates to the higher of the reference projections. In the case of Western Europe, the IWP forecast is for growth of at least 1.2 percent, compared with *FAO* and *CHASE scenario* projections of 0.6 to 1.0 percent growth. For the USA, the projected growth of 0.2 to 0.5 percent is somewhat lower than the 0.7 to 0.9 percent forecast by the IWP.

The IWP indicated alternative forecasts for consumption for *newsprint, printing and writing* and *other paper and paperboard* for a number of countries. In TABLE 29 these forecasts are totalled for the included countries and compared with the projections.

Total paper and paperboard consumption

This section sums up the discussion of consumption of *total paper and paperboard*, the projections for which are shown in FIGURE 23 and TABLES 27 and 28.

Table 27 - TOTAL PAPER AND PAPERBOARD CONSUMPTION PROJECTIONS TO 1995 (million metric tonnes)

FAO regions and subregions	Actual 1984	FAO		CHASE	
		1990	1995	1990	1995
World	187.3	215.9	254.8	214.0	245.9
Developed market economies	144.4	156.1	177.1	155.9	172.7
North America	73.4	76.1	82.6	78.3	84.6
West Europe	46.7	50.6	58.9	50.4	56.8
Oceania	2.5	2.8	3.2	2.9	3.3
Others	21.6	26.5	32.2	24.2	27.9
Developing market economies	20.1	27.9	37.0	26.5	34.7
Africa	8	1.1	1.4	1.1	1.3
Latin America	9.7	13.3	17.0	11.9	15.2
Near East Africa	4	7	9	6	7
Near East Asia	1.3	1.7	2.2	1.6	2.1
Far East	7.7	10.9	15.2	11.1	15.2
Centrally planned economies	22.8	31.8	40.6	31.5	38.5
Asia	7.5	12.1	16.2	14.5	20.0
Europe and USSR	15.2	19.7	24.4	16.9	18.4

30

Table 28 - PAST AND PROJECTED ANNUAL GROWTH IN CONSUMPTION (percent per year)

	Actual	FAO		CHASE	
	1970-82*	1982*-90	1990-95	1982*-90	1990-95
World	2.7	2.8	3.4	2.7	2.8
Developed market economies	2.2	2.1	2.6	2.1	2.1
North America	1.9	1.6	1.6	2.0	1.6
West Europe	2.1	2.0	3.1	1.9	2.4
Oceania	2.5	1.7	3.0	2.4	2.3
Others	3.1	3.7	4.0	2.6	2.9
Developing market economies	6.0	5.0	5.8	4.3	5.5
Africa	3.7	3.9	4.9	3.5	4.0
Latin America	4.8	4.3	5.1	3.0	4.9
Near East Africa	6.2	6.3	5.7	4.3	3.3
Near East Asia	7.0	3.5	5.3	3.1	4.8
Far East	8.2	6.2	6.8	6.4	6.5
Centrally planned economies	3.9	4.6	5.0	4.5	4.1
Asia	7.4	6.4	6.0	8.9	6.6
Europe and USSR	2.6	3.6	4.4	1.7	1.7

82* = 1980-1984 average

Table 29 - TOTAL PAPER AND PAPERBOARD CONSUMPTION IN 1995 Comparison of IWP forecasts with projections (million metric tonnes)

	1995		
	Projection		Forecast
Product	FAO	CHASE	IWP
Newsprint	30.3	29.8	27.0
Printing and writing	59.3	57.6	51.5
Other paper	95.5	92.5	97.3 to 104.2
Total* (selected countries)	185.1	179.9	175.8 to 182.7
World total	254.8	245.9	240.8 to 250.3

* Comparison made only for those countries for which IWP made forecasts.

Table 30 - PROJECTED PAPER AND PAPERBOARD CONSUMPTION Statistical confidence limits (million metric tonnes)

Range	1995
Lower limit — 90 percent confidence	236
Most expected consumption CHASE scenario	246
Most expected consumption FAO scenario	255
Upper limit 90 percent confidence	265

In the discussion of each product the statistical uncertainty associated with the projections is described. The confidence limits for the projection of total paper consumption are ±4 percent (FIGURE 24). The projected range within which total consumption is expected to fall in 1995 is shown in TABLE 30.

A further source of uncertainty stems from the economic forecasts. This is illustrated by the difference between the projections obtained under the FAO and CHASE scenarios. The result of sensitivity analysis (FIGURE 25) in which the annual rate of growth in all years is increased or decreased by 1 percent is shown in TABLE 31.

Table 31 - PROJECTED PAPER AND PAPERBOARD CONSUMPTION Sensitivity to 1 percent change in economic growth (million metric tonnes)

	1995		
Scenario	−1 percent	Most expected	+1 percent
CHASE	221	246	275
FAO	228	255	285

Figure 23

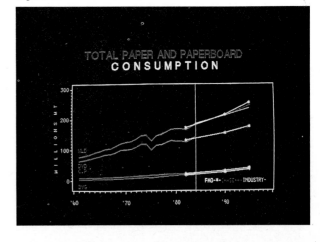

Figure 24

Figure 25

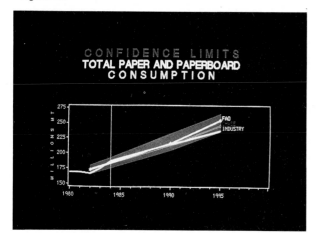

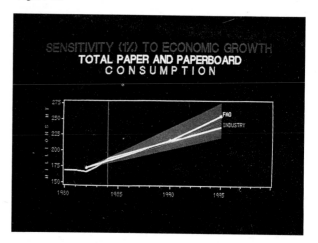

CHAPTER 4

SUPPLY OF PAPER AND PAPERBOARD

This chapter discusses production of paper and paperboard and distribution, and explores the possibilities of projecting future development. As the information available relating to production and capacity and supply development is limited, the results presented should be regarded as a tentative first step in developing an international modelling approach to this aspect of the sector's development.

Production and distribution

World total production of paper in 1984 was 187 million tonnes. The average growth in production in the decade of the 1970s was 2.7 percent. Growth in the *developed market* and *centrally planned countries* was approximately equal to growth in their consumption. Paper is produced in some 90 countries (FIGURE 26). Production is however, very concentrated: three countries — United States, Japan and Canada — account for more than 50 percent of world production. The 12 countries with more than 3 million tonnes annual production account for 80 percent of world production, and 25 countries with over 1 000 000 tonnes account for more than 90 percent.

Many countries, mainly relatively small consumers, are wholly dependent on imports to meet their consumption requirements and most producing countries are involved in trade, but a relatively small number of countries are mainly involved in the production of paper for export rather than for domestic production. Predominant among the countries producing mainly for export are Canada, Finland and Sweden, with 70-80 percent of their production going to export.

Of the major categories of paper, the production of newsprint is the most highly concentrated. Five countries account for 70 percent of world production and Canada, Finland and Sweden export 90 percent of their production contributing 80 percent of world trade volume of this product. Production of *printing and writing papers* and *other paper and paperboard* is more widely dispersed as in the trade. Finland, however, leads in export of *printing and writing paper*, and Sweden in *other paper and paperboard*. Very few developing countries have entered significantly into the business of production of paper for export. Brazil is the most notable with exports in 1984 of 700 000 tonnes.

Several developing countries are involved in production of pulp for export — pulp exports of Brazil, Chile and Swaziland exceed 100 000 tonnes per year. Several developed countries are also major net exporters of pulp. The discussion of pulp production is taken further in Chapter 5.

Building on the expert contributions from the Industry Working Party, factors which may have particular influence on the development of supply are reviewed qualitatively. Then, as was done with demand, projections based on analysis of the relationship between past production and economic variables are presented. The IWP considers that *FAO Pulp and Paper Capacities* provides a valuable indication of the upper limit on the development of production to 1990. Projections of production are reviewed in this sense.

Factors influencing supply development

The analysis of past development of production indicates that the industry has been able to expand its investment and production to meet the growth in consumption at prices which have perhaps had a declining trend, but which have not changed dramatically in real terms. This has been achieved in spite

Figure 26

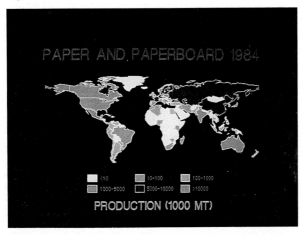

PAPER AND PAPERBOARD 1984

PRODUCTION (1000 MT)

of apparently adverse factors such as increasingly strict environmental controls, the higher cost of energy and fibre supplies.

Contributing to this success has been the application of improved processes by the industry, and its responsiveness to the changing requirements of the uses of paper and paperboard. For example, thermomechanical pulping (introduced in the early 1970s) and more recently chemi-thermomechanical pulping have both enabled reductions in usage of more costly chemical pulp as well as in wood consumption. The industry has further improved its economy in wood utilization by increasing use of sawmill residues and of waste paper. The long-standing practice of recovering chemicals and energy from the combustion of pulping waste liquors, together with the increased use of wood residues as fuel for power generation, have greatly reduced the industry's dependence on purchased energy. While some mills even have surplus energy for sale, the industry as a whole remains a major net energy buyer.

The industry has responded to a need for paper of lighter basis weight while meeting strength and quality requirements. This is most noticeable in the case of newsprint but has been important in printing and packaging papers. The industry has also responded to the diverse requirements of development in printing technology, and in copying and office machinery. We see the assault that paper has made, in sophisticated combination with other materials, on the exacting requirements of packaging for an increasing variety of powders and fluids, foodstuffs and beverages, in a way that satisfies requirements for protection and hygiene and, at the same time, remains cost effective.

In reviewing the future development of supply, the IWP considered four broad areas to be of critical importance: fibre resources; the availability of labour, energy and chemicals; capital; and necessary physical and human infrastructure.

Fibre resources

For the forest rich areas of North America, and the USSR with substantial coniferous resources and for tropical countries of South America, Southeast Asia and parts of Africa with tropical forest resources and land for fast growing coniferous and deciduous plantations, there is a fibre resource potential for industry expansion (FIGURE 27). In a number of European countries the potential for expansion is limited by the forest resources available while in others, such as France, Portugal and Spain recent plantation programmes have established substantial potential. Similar potential for plantation supplies exists in Australia and in New Zealand in the later 1990s.

The effects on future forest production from

Figure 27

atmospheric pollution is, however, a source of uncertainty in a number of northern countries.

There are a number of countries where fibre supply is a critical factor in industry expansion due to limited forest resources and intense competition for wood and other fibre for other uses: examples are China and the countries of South Asia. In several countries baggasse is the raw material for pulp production and future supply is made uncertain by the recession in sugar production for export. Finally there is a group of countries whose production will be largely dependent on the availability of fibre imports.

The location of wood raw material is frequently less than ideal in relation to the optimum location for the industry. In Australia, for example, the industry may be forced to locate near the coast to secure adequate water supplies while the forest is located some distance inland. The optimum wood supply areas are in many counties located at considerable distance from the centres of the domestic market and other industrial infrastructure.

Other inputs

Availability of chemical inputs will not be a critical factor in determining development. Though labour is not considered likely to be a constraining factor in terms of numbers, technical ability of the labour force is an important consideration. Where pulping process and scale permits, the availability of low cost energy through recycling within the industry may lead to comparative advantage in some countries. The movement toward a higher proportion of paper manufacture in exporting countries which have to import their energy supplies may be costly in a situation where high energy prices prevail. The relatively high wood, labour and energy costs in several of the larger producing countries in Europe lead those producers to concentrate on high value final products.

Capital

The industry has been operating in a period of low profitability in many countries, and has thus not been generating interal capital resources. The uncertainties relating to the volatility of exchange rates are unfavourable where foreign markets o foreign capital are concerned. Problems of balance of payments and debt servicing are also significant constraints on the mobilisation of foreign capital. In certain countries the heavy dependence of the economy on export of a few basic commodities contributed a further element of uncertainty for the potential investor. There are advantages for countries where domestic capital and locally produced equipment can be mobilised.

Infrastructure

The difficulty of investment in such a capital intensive industry is related to the state of the supporting infrastructure. Availability of sites with the physical requirements necessary for the mill may be a limitation, and the availability of locations where environmental requirements, can be met may be even more restricted.

Essential requirements are the transport connections, power supplies, communications system and the community infrastructure. If these are not already established, they become an essential component of the investment, and an addition to the cost of generating supply.

The efficient operation of the industry requires staff of appropriate technical and managerial skills; if not available within the community, this staff has to be brought in from outside. For the efficient running of an industry of any significant dimensions, the support of education and training for the broad range of knowledge and skills necessary for management and operation has to be available. Once these facilities are established and the skills developed, the skills themselves may become a valuable source of export earnings. Securing this support is an essential component of the investment; if it is not available within the country effective linkages to outside are vital. The industry also must have access to research and development, and adequate technical and market information.

Projected production of paper and paperboard

Projected world production is equal to projected consumption. As mentioned in Chapter 7 on methodology, the projection of production is based on a model of supply which relates past paper production to the economic indicator of investment activity (gross fixed capital formation) and the price of a major input (pulp). It is recognised that this model may not be sensitive to important factors effecting the comparative advantage of producers in different countries and regions or their preference for investment in pulp and paper compared with other sectors. As with the projections of consumption, those for supply must be regarded as reference projections based on the assumption that the past relationships of the economic variable to the forecast will continue.

The projections for *total paper and paperboard* are shown in TABLES 32 and 33. World production of *paper and paperboard* is projected to increase from

Table 32 - TOTAL PAPER AND PAPERBOARD PRODUCTION PROJECTED TO 1995 (million metric tonnes)

FAO regions and subregions	ACTUAL 1984	FAO		CHASE	
		1990	1995	1990	1995
World	187.6	215.9	254.8	214.0	245.9
Developed market economies	150.2	169.5	196.4	168.8	190.6
North America	76.5	83.5	93.2	87.7	97.0
West Europe	49.9	56.3	66.1	54.0	62.0
Oceania	2.2	2.4	2.9	2.3	2.6
Others	21.4	27.2	34.0	24.6	28.8
Developing market economies	15.3	17.2	22.1	16.5	21.3
Africa	3	4	5	4	5
Latin America	8.6	10.0	12.5	7.9	10.1
Near East Africa	1	1	1	1	2
Near East Asia	6	7	1.0	8	1.0
Far East	5.5	5.8	7.8	7.1	9.4
Centrally planned economies	22.0	29.1	36.2	28.6	33.8
Asia	7.0	10.5	13.7	12.3	16.2
Europe and USSR	15.0	18.5	22.5	16.3	17.6

Table 33 - PAST AND PROJECTED ANNUAL GROWTH IN PRODUCTION (percent per year)

	ACTUAL	FAO		CHASE	
	1970-82*	1982*-90	1990-95	1982*-90	1990-95
World	2.7	2.7	3.4	2.6	2.8
Developed market economies	2.2	2.5	3.0	2.4	2.5
North America	1.9	2.0	2.2	2.6	2.0
West Europe	2.3	2.6	3.3	2.1	2.8
Oceania	3.0	1.8	3.3	1.2	2.2
Others	3.1	4.1	4.6	2.8	3.2
Developing market economies	7.8	2.9	5.1	2.3	5.3
Africa	6.6	1.8	4.8	2.1	4.0
Latin America	6.7	3.0	4.5	0.0	5.0
Near East Africa	0.5	0.9	6.2	4.3	4.0
Near East Asia	10.0	2.8	5.1	3.2	4.7
Far East	10.2	2.7	6.0	5.2	5.8
Centrally planned economies	3.8	3.9	4.5	3.7	3.4
Asia	7.0	5.4	5.4	7.5	5.7
Europe and USSR	2.7	3.1	3.9	1.5	1.6

82* = 1980-1984 average

187.6 million tonnes in 1984 to 254.8 million tonnes under the *FAO scenario* or 245.9 million tonnes under the *CHASE scenario*, an increase in the range of 58 to 67 million tonnes. Some 40-46 million tonnes is the projected increase in *developed market economies*, 6-7 million tonnes in *developing market economies* and 11-14 million tonnes in *centrally planned economies*. Projections for the total and for main products are shown in FIGURES 28 to 31.

Deriving from the forecasts of capital formation, the projected growth rates of production are slightly above the rates in the past decade for *developed market economies* and *centrally planned economies* but lower for the *developing market economies*. The projected growth of production for *developed market economies* is also slightly above their projected growth for consumption, whereas for the other regions it is below consumption growth.

Figure 29

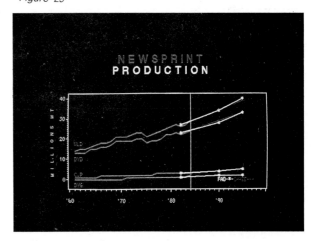

Figure 28

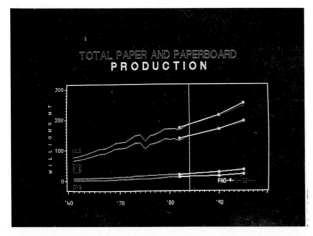

Figure 30

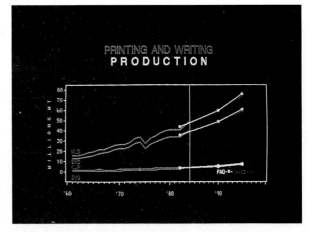

Figure 31

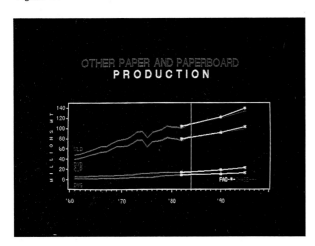

Figure 33

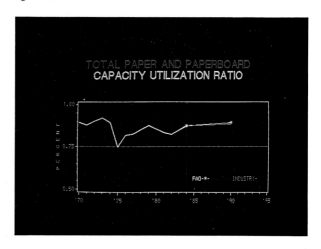

Production and capacity

Projections of production to 1990 are compared with the estimates of capacity development provided by *FAO Pulp and Paper Capacities*. This capacity data has the limitation that for a number of countries the estimates have not been updated to 1990. In any case, all capacity surveys tend to underestimate capacity additions beyond 2 to 3 years from the survey date. The comparison shows that for total paper the *FAO and CHASE scenario* production projections to 1990 of 215.9 and 214 million tonnes are 13-15 million tonnes lower than the capacity predicted in the survey to 1990 for the world (FIGURE 32). The projections are lower for all regions except the *centrally planned economies*. For *newsprint*, the 1990 CHASE projection of world production exceeds 1990 capacity. This reflects predicted 1990 capacity lower than 1990 production in the *developed* and in the *centrally planned countries*.

Relating production to capacity, worldwide capacity utilisation in 1984 was 88 percent, reflecting 91 percent utilization in *developed market econ-*

Figure 32

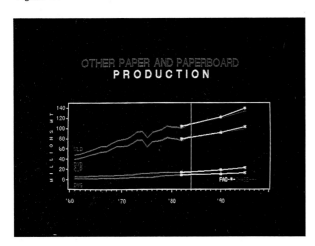

Table 34 - **PROJECTED TOTAL PRODUCTION, 1990 (CHASE)** Compared with predicted capacity for 1990 (million metric tonnes)

Product and Projection	Developed market economies	Developing market economies	Centrally planned economies	World Total
TOTAL PAPER AND PAPERBOARD				
Production 1984	150.2	15.3	22.0	187.6
Capacity 1984	165.7	20.1	26.5	212.4
Production 1990				
FAO scenario	169.5	17.2	29.1	215.9
CHASE scenario	168.8	16.5	28.6	214.0
Capacity 1990	178.2	23.3	27.2	228.8
NEWSPRINT				
CHASE scenario	28.7	1.6	3.8	34.2
Capacity 1989	27.6	2.7	2.9	33.3
PRINTING AND WRITING				
Production 1990				
CHASE scenario	48.7	4.7	6.4	59.9
Capacity 1990	50.6	5.8	5.1	61.6
OTHER PAPER AND PAPERBOARD				
Production 1990				
CHASE scenario	91.3	10.1	18.3	119.7
Capacity 1990	99.9	14.7	19.1	133.8

omies, 76 percent in *developing* and 83 percent in *centrally planned economies*. The equivalent ratios of projected production in 1990 to predicted capacity 1990 are 94, 95, 74 and 107 percent repectively (FIGURE 33). The production projected for *centrally planned economies* exceeds predicted capacity (TABLE 34).

Examination of the capacity forecast with individual country production projections (TABLE 35) shows that the capacity prediction for many countries lies in the same range as the production projections to 1990. The projected production to 1990 lies within the capacity predicted to 1990, except in the case of a small number of larger producers where either the *FAO* or *CHASE scenario* projection exceeds the capacity prediction. These are the United States, which is discussed more fully below, Australia, Belgium, France, Republic of Korea and

Table 35 - TOTAL PAPER AND PAPERBOARD (1 000 metric tonnes)

COUNTRY	1984		1990		
	CAPACITY	PRODUCTION	ESTIMATED CAPACITY	PROJECTED PRODUCTION	
				FAO	CHASE
Argentina	1 100	948	1 278	793	1 060
Australia	1 586	1 518	1 712	1 564	1 731
Austria	2 198	1 920	2 447	2 094	2 079
Bangladesh	149	150	120	114	116
Belgium-Lux	912	912	1 201	919	1 289
Brazil	4 228	3 768	5 318	3 727	4 703
Canada	15 548	14 220	17 149	15 373	15 564
Chile	411	378	605	394	402
China	7 805	7 560	12 250	12 137	10 380
Colombia	571	366	665	454	475
Costa Rica	19	12	19	32	32
Cuba	158	108	258	123	122
Czechoslovakia	1 364	1 236	1 560	1 348	1 361
Denmark	375	330	390	408	473
Dominican Rep.	64	12	64	10	10
Egypt	205	108	241	171	124
El Salvador	56	18	56	30	30
Ethiopia	11	12	31	10	10
Finland	7 865	7 320	9 135	7 868	7 885
France	5 707	5 568	6 702	5 822	6 677
Germany, Fed. Rep	9 865	9 162	12 500	9 734	10 033
Greece	508	276	508	335	407
Guatemala	46	18	46	37	37
Hungary	506	504	558	553	746
Israel	161	150	161	208	207
Italy	5 950	4 722	5 950	5 291	5 662
Japan	23 495	19 344	25 365	22 963	25 273
Jordan	8	6	17	5	5
Kenya	80	72	95	80	84
Korea, Rep. of	2 456	2 208	2 797	3 018	2 373
Malaysia	80	78	505	87	47
Mexico	3 096	2 190	3 704	1 625	2 193
New Zealand	1 308	732	1 488	809	760
Nigeria	15	18	225	27	18
Norway	1 815	1 560	1 860	1 665	1 637
Panama	24	42	26	44	43
Paraguay	7	12	9	14	14
Philippines	476	252	457	152	296
Poland	1 373	1 254	1 650	1 375	1 523
Portugal	720	588	770	573	590
South Africa	1 615	1 962	2 113	1 466	1 729
Spain	3 290	2 952	3 345	3 233	3 173
Sudan	8	12	8	11	11
Sweden	7 375	6 870	8 835	7 171	7 220
Switzerland	1 050	984	1 075	1 069	1 078
Thailand	558	306	730	407	472
Tunisia	50	24	92	26	26
United Kingdom	3 766	3 588	4 576	4 755	4 529
Uruguay	90	48	90	32	53
USA	67 287	62 364	72 840	72 372	67 961
Venezuela	751	558	797	407	615

Note: FAO data from "Pulp and Paper Capacities 1984-1989 and 1985-1990" and "Yearbook of Forest Products 1984".

the United Kingdom. In the last case both projections are higher than predicted capacity. Possibly the tendency for the capacity estimate to be lower can be attributed to the fact that it relates to an earlier year. The prediction capacity for the United States is compared with the production projections in TABLE 36.

A general conclusion from this review is that with the exceptions indicated, the projected production is not significantly in conflict with the capacity prediction provided by the capacity survey. It has however, to be borne in mind that capacity utilisation in major producing countries is in the range of 85-95 percent, while it is frequently at a lower level in smaller developing countries.

Table 36 — COMPARISON OF CAPACITY FORECAST AND PRODUCTION PROJECTIONS FOR USA IN 1990
(million metric tonnes)

Product	FAO capacity forecast	Production FAO	Projections CHASE
Newsprint	5.2	6.3	8.1
Printing and writing	19.8	20.0	21.2
Other	45.6	41.6	42.8
Total	70.8	67.9	72.3

CHAPTER 5

DERIVED DEMAND FOR INPUT

The production of paper involves the use of wood pulp, other fibre pulp, waste paper and additives such as pigments, fillers, coatings, binders, sizes and starch. Collectively, they are called the furnish. The furnish consumed to make the 187 million tonnes of paper in 1984 is shown in TABLE 37. FIGURE 34 illustrates the flows comprising the fibre balance.

Figure 34

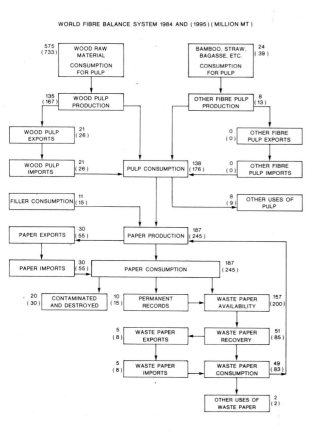

WORLD FIBRE BALANCE SYSTEM 1984 AND (1995) (MILLION MT)

Furnish composition

Wood pulp contributes 65 percent of the total furnish. The main types of woodpulp are chemical, mechanical, thermomechemical, chemi-thermo-mechanical, and semi-chemical.

Mechanical pulps are produced by grinding wood billets, or refining wood chips. Thermomechanical and chemi-thermomechanical processes refine chips which have been softened by heating under pressure (thermomechanical) or by a combination of heat and chemical treatment (chemi-thermomechanical). A characteristic of mechanical pulps is the high yield of pulp produced from wood consumed, typically over 90 percent. Papers which contain them are often described as "wood containing".

Semi-chemical pulps are also made by a combination of chemical and mechanical treatment. This type of pulp manufacture is usually integrated with the production of fluting medium.

Chemical pulps are divided into two major groups: sulphite pulps in which wood chips are cooked under pressure in a bisulphite liquor (ammonium, calcium, magnesium or sodium-base), and sulphate pulps for which the cooking liquor includes sodium hydroxide (soda pulp) or sodium sulphide (sulphate or kraft pulp). The yield of chemical pulp is much lower than mechanical, typically 40-50 percent. Wood pulp is made from wood of both coniferous and nonconiferous wood species and may be bleached or unbleached.

Pulp properties which are important in papermaking are determined by a combination of the nature of the wood, the type of pulping process and the pulp yield. Any type may be used in papers and paperboards as the sole furnish component or in

Table 37 - FURNISH COMPOSITION OF TOTAL PAPER AND PAPERBOARD PRODUCTION IN 1984 (million metric tonnes)

Type of Economy	Paper production	Apparent consumption				
		Total furnish	Wood pulp	Other fibre	Waste paper	Additives
Developed market	150	159	109	1	40	9
Developing market	15	16.5	7	3	6	0.5
Centrally planned	22	23.5	15	4	3	1.5
World total	187	198	130	8	49	11

blends with other wood pulps, waste paper and additives. The composition of the 130 million tonnes of world wood pulp production in 1984 is indicated TABLE 38.

Table 38 - COMPOSITION OF WORLD WOOD PULP PRODUCTION IN 1984 (percent)

	WOOD SPECIES		
	Coniferous	Non-coniferous	Total
Mechanical	16	1	17
Thermo- and chemi-thermochemical	8		8
Semichemical	1	6	7
Sulphite	7	1	8
Sulphate plus soda	40	20	60
bleached	18	17	35
unbleached	22	3	25
Total	72	28	100

Eight million tonnes, 4 percent of the world total furnish, consists of other fibre pulp. This type of pulp contributes 20 percent of the furnish for developing country paper production. (It is the main type currently produced in China.) Important raw materials are straw, bagasse and bamboo.

Waste paper accounts for 25 percent of the total furnish. Waste paper's share total furnish has increased strongly from about 18 percent in 1970. It varies between countries. For several countries — Japan, United Kingdom, Federal Republic of Germany, Republic of Korea and Indonesia — it constitutes around 50 percent of the total furnish. The orignal paper type of the wastepaper determines the type of pulp that it can replace. Several developing countries with small paper industries depend largely on waste paper and pulp imports. Major paper exporting countries such as Canada, Finland and Sweden have less than 10 percent of waste paper in their furnish.

Non-fibrous additives such as coating pigments and binders, fillers, opacifiers, dyestuffs, optical bleaching agents, sizes and strength and functional additives are widely used in many types of papers and paperboards. The most important additives in terms of tonnage are coating pigments and fillers. World consumption of these in 1982 has been estimated at 11 million tonnes (Galsworthy, 1984). The most commonly used material is kaolin, accounting for 80 percent of the market, but calcium carbonate, which is cheaper, is also used for papers produced under neutral or alkaline conditions. About 75 percent of the total coating pigment and filler usage is in printing and writing papers to obtain the required level of characteristics such as surface finish, opacity and printability — especially for colour printing. The use of such materials can also be advantageous because they are cheaper on a weight basis than the pulps with which they are used.

Figure 35

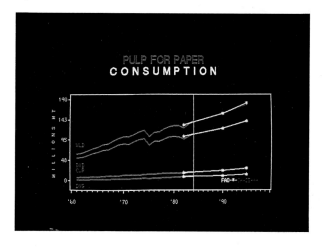

Estimating future furnish

Two approaches are adopted to projecting pulp, waste paper and fillers consumption. The first is a statistical approach and the second is the estimation of detailed composition of fibre furnish by paper grade and its development, by the Industry Working Party. In the statistical approach the past consumption of pulp and waste paper is related to the production of the main paper grades. The statistical relationship is used with the projected production of paper to estimate future pulp and waste paper consumption. The projection is of total pulp by main categories, and waste paper for total paper production. This approach allows the projection of pulp and waste paper consumption for each paper producing country in relation to its projected paper production. The lack of systematic statistics on consumption of additives prevents the inclusion of projections of the consumption of fillers, coating and pigments.

The projections (TABLE 39) indicate an increase in pulp consumption from 140 million tonnes to 177 to 181 million tonnes (FIGURE 35). The ratio of pulp per ton of paper would fall from 0.74 to 0.72 tonnes

Figure 36

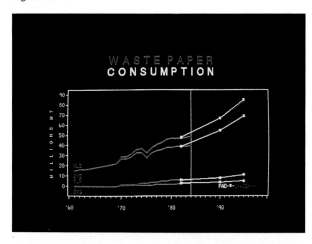

Table 39 - PROJECTED CONSUMPTION OF FIBRE FOR PAPER PRODUCTION IN 1995 Under alternative scenarios

Type of Economy	Paper production	Total fibre	Mechanical pulp	Chemical pulp	Other fibre	Waste paper
Developed market						
1984	150.2	150.2	25.9	83.1	0.8	40.2
1995 FAO	196.4	207.8	29.0	107.8	1.6	69.2
CHASE	190.6	204.6	28.8	106.5	1.6	67.5
Developing market						
1984	15.3	17.0	1.1	6.3	3.3	6.2
1995 FAO	17.2	25.1	1.1	9.1	3.8	11.0
CHASE	21.3	24.4	1.1	8.3	4.1	10.7
Centrally planned						
1984	22.0	21.5	2.9	11.5	4.2	2.7
1995 FAO	36.2	33.5	3.3	17.9	7.1	5.0
CHASE	33.8	31.3	2.9	15.3	7.9	14.9
World total						
1984	187.6	188.8	29.9	101.1	8.4	49.1
1995 FAO	254.8	266.5	33.5	134.9	12.6	85.3
CHASE	245.9	266.5	32.9	130.3	13.7	83.3

per ton of paper. Waste paper consumption is projected to increase sharply from 49 million tonnes to 83 to 85 million tonnes, or from 0.26 to 0.33 tonnes per ton of paper by 1995 (FIGURE 36).

Projected pulp consumption and capacity

The consumption of pulp for paper production projected for 1990 (TABLE 40) is within the volume of predicted capacity from *FAO Pulp and Paper Capacities* for 1990. The FAO capacity survey allocates a higher volume to mechanical, thermomechanical, chemical and semi-chemical, but a slightly lower volume to other fibre pulp.

Table 40 - PROJECTED PULP CONSUMPTION COMPARED WITH PREDICTED CAPACITY IN 1990 (million metric tonnes)

Product	Capacity 1990 FAO	Projected consumption 1990 FAO	Projected consumption 1990 CHASE
Mechanical including thermomechanical	39.3	30.7	30.8
Chemical including semi-chemical	116.7	115.8	115.1
Other fibre pulp	10.4	10.3	11.2
Total pulp for paper	166.4	156.9	157.2

Roundwood requirements

The projection of industrial roundwood for all wood using industries indicates an increase in world consumption from 1 240 million m^3 in 1984 to 1 600 million or 1 540 million m^3 in 1995. Separation of pulp wood in the projection was not possible. Using very general consumption ratios of 2.5 m^3 of wood per ton of mechanical pulp and 5 m^3 per ton of chemical pulp, the estimated wood input for paper pulp of 575 million m^3 in 1984 increases to 758 million or 733 million m^3 in 1995. The regional composition of the lower projection is shown TABLE 41.

Table 41 - PROJECTED CONSUMPTION OF INDUSTRIAL ROUNDWOOD AND PULPWOOD (million cubic metres)

Type of Economy	1984 ACTUAL Total industrial wood	1984 ACTUAL Pulp-wood	1995 CHASE Total industrial wood	1995 CHASE Pulp-wood
Developed	774	479	938	605
Developing	179	33	205	45
Centrally planned	289	62	354	83
Total world	1 244	575	1 540	733

42

Fibre furnish by paper grade

The IWP reviewed the consumption of components of the fibre furnish by individual paper grades. A survey conducted through the IWP yielded responses from 25 countries and was used to estimate the furnish composition for regional groupings of countries, and for making judgements on prospective changes in furnish composition (TABLES 42 and 43). Also the IWP reviewed outlook for the furnish composition for each paper product.

Newsprint (production level 30 million tonnes)

Formerly, newsprint furnish consisted of 80 percent groundwood plus 20 percent bleached sulphite pulp. Recently thermomechanical (TMP) and chemithermomechanical (CTMP) pulps have provided a larger share of the products furnish. Sulphite pulp is being replaced by bleached sulphate. Sulphate pulp may in turn, be replaced by CTMP. In several countries newsprint is being produced by recycling de-inked newspapers.

Coated wood-containing paper (production level 7 million tonnes)

This is mainly a product of Europe and the United States. Composition (for 1 000 kg of paper) is typically 350-400 kg of mechanical pulp, 350-400 kg of chemical pulp, and 300-350 kg of fillers and pigments. The USA uses somewhat lower filler and pigment volume. The use of CTMP/TMP in the furnish is expected to increase.

Coated wood-free paper (production level 4 million tonnes)

This is mainly produced in the United States and Western Europe. The furnish is 750 kg of chemical pulp plus 300-400 kg of fillers and pigment. Of the chemical pulp, more than half (350-400 kg) is bleached hardwood. The use of bleached hardwood and fillers may increase.

Table 42 - AVERAGE FURNISH OF TOTAL PAPER AND PAPERBOARD PRODUCTION
(kilograms required per 1 000 kilograms of paper)

| | | PAPER PRODUCTION | TOTAL FURNISH | FURNISH | | | | |
| | | | | Filler and pigment | Total fibre | FIBRE | | |
						Wood pulp	Waste paper	Other fibre
Algeria	1984	1 000	1 102	37	1 065	556	509	—
Argentina	1983	1 000	1 103	40	1 063	625	347	92
Australia	1983	1 000	1 020	32	988	668	313	7
Brazil	1983	1 000	1 032	43	989	661	294	35
Canada	1982	1 000	1 038	—	1 038	955	82	1
Chile	1983	1 000	1 253	37	1 216	938	139	139
China - Taiwan	1983	1 000	1 060	83	977	218	194	565
Czechoslovakia	1983	1 000	1 086	56	1 029	719	308	2
El Salvador	1984	1 000	1 042	—	1 042	417	625	—
Finland	1983	1 000	1 046	117	929	883	44	2
France	1982 bis	1 000	1 000	91	909	594	278	37
France	1983	1 000	1 217	167	1 050	617	426	7
Germany, Fed. Rep.	1982 bis	1 000	1 000	126	874	507	362	4
Indonesia	1983	1 000	1 110	80	1 029	528	499	3
Indonesia	1984	1 000	977	7	970	650	—	320
Italy	1982 bis	1 000	1 000	44	956	436	475	45
Japan	1984	1 000	1 059	—	1 059	551	506	2
Korea, Rep. of	1984	1 000	1 005	—	1 005	363	638	3
Morocco	1984	1 000	1 000	—	1 000	474	526	—
New Zealand	1983	1 000	1 000	7	993	886	107	—
Panama	1984	1 000	1 000	—	1 000	105	895	—
Philippines	1983	1 000	1 059	3	1 055	630	388	38
Portugal	1983	1 000	1 050	32	1 018	651	366	2
Spain	1982 bis	1 000	1 000	12	988	502	434	51
Sweden	1983	1 000	1 073	55	1 018	913	105	—
Switzerland	1982 bis	1 000	1 000	54	946	555	361	30
Tunisia	1984	1 000	1 000	—	1 000	500	194	306
United Kingdom	1982 bis	1 000	1 000	59	941	444	475	22
	1983	1 000	1 014	—	1 014	442	562	10
USA	1983	1 000	1 000	—	1 000	753	244	2

Table 43 - WOOD PULP COMPONENT OF AVERAGE FURNISH FOR TOTAL PAPER AND PAPERBOARD PRODUCTION (kilograms per 1 000 kilograms of paper)

		TOTAL WOOD PULP	MECHANICAL	THERMO-MECH.	SEMI-CHEMICAL	WOOD PULPS — CHEMICAL — Total chemical	SULPHITE Unbleached	SULPHITE Bleached	SULPHATE Unbleached	SULPHATE Bleached coniferous	SULPHATE Bleached non-coniferous
Algeria	1984	556	9	—	—	546	—	—	269	278	—
Argentina	1983	625	166	—	37	422	—	33	152	88	149
Australia	1983	668	105	166	52	345	90	22	76	69	83
Brazil	1983	661	59	—	39	563	2	6	218	310	26
Canada	1982	955	589	—	23	343	121	3	135	38	51
Chile	1983	938	528	—	—	410	—	77	139	191	—
China - Taiwan	1983	218	53	5	—	160	11	9	99	47	—
Czechoslovakia	1983	719	131	—	73	515	114	80	165	156	—
El Salvador	1984	417	42	—	—	375	—	—	—	375	—
Finland	1983	883	274	108	52	449	12	14	146	163	114
France	1982 bis	594	91	—	27	477	6	40	90	144	197
France	1983	617	93	—	26	497	2	33	104	358	—
Germany, Fed. Rep.	1982 bis	507	163	—	10	334	5	86	8	87	149
Indonesia	1983	528	—	13	54	461	—	—	—	408	—
Indonesia	1984	650	—	15	120	515	—	2	—	505	—
Italy	1982 bis	436	117	—	25	293	9	52	19	69	143
Japan	1984	551	132	—	—	419	—	7	99	313*	—
Korea, Rep. of	1984	363	77	3	8	275	—	6	77	192	—
Morocco	1984	474	21	—	—	454	—	10	82	361	—
New Zealand	1983	886	399	—	7	480	—	—	295	141	44
Panama	1984	105	—	—	—	105	—	53	—	53	—
Philippines	1983	630	111	38	—	481	—	—	235	246	—
Portugal	1983	651	—	—	—	651	—	2	378	273	—
Spain	1982 bis	502	71	—	15	417	1	13	118	176	108
Sweden	1983	913	146	105	39	622	27	41	323	146	86
Switzerland	1982 bis	555	216	—	7	331	9	98	17	79	129
Tunisia	1984	500	28	—	—	472	—	28	139	306	—
United Kingdom	1982 bis	444	55	—	22	368	8	32	16	149	162
	1983	442	56	—	30	357	8	28	18	140	161
USA	1983	753	46	30	60	618	30	—	298	290	—

* Includes non-coniferous.

Uncoated wood-containing paper (production level 10 million tonnes)

The principal producers are in Europe, North America and Japan. The furnish typically consists of 550-650 kg of mechanical pulps, 250-350 kg of chemical pulps and 100-150 kg of fillers and pigments. In Western Europe and USA, 120-150 kg of fibre is in the form of waste paper. The use of CTMP is increasing in all regions because of the comparative advantage of the technology where low cost energy is available, or where spruce and aspen are available.

Uncoated wood-free paper (production level 24 million tonnes)

Production of this grade is the most widespread. Although North America, Europe and Japan predominate, it is also important in all other regions. The predominant component in most regions is chemical wood pulp, with around 800 kg per ton of paper. The main variation is in the component of bleached hardwood pulp, which ranges from 400 kg in the northern hemisphere to 700 kg in the southern hemisphere, where eucalyptus is the pulpwood species. In India and China it is estimated that non-wood fibre pulps play the major role with around 600 kg per ton.

In Europe, the pigment and filler content is higher and is likely to persist in view of optical properties required, and the cost pressures which tend to favour substitution of filler for fibre.

The main trends are expected to be toward increased use of bleached hardwood sulphate pulp and reduced use of sulphite and bleached coniferous sulphate pulps. Changes in technology toward neutral or alkaline papermaking will permit wider use of the cheaper calcium carbonate and will favour increased use of pigments and fillers. It is likely that the main growth in countries currently using other fibre will be toward increased use of wood pulp.

Household and sanitary papers (production level 10 million tonnes)

The main pulp component is chemical wood pulp ranging from 450-750 kg, with the exception of China where it is estimated to be mainly pulp of other fibre. There is generally a mechanical pulp component of 50-100 kg while the waste paper component ranges from 250-400 kg. The high volume grades tend to be based predominantly on waste, while quality grades use mainly bleached sulphate pulps. In the sulphate component, hardwood pulps predominate in Latin America.

Trends will be toward increased use of thermomechanical pulps bleached hardwood sulphate pulp and waste paper, and decreased use of mechanical and coniferous sulphate pulps.

Linerboard (production level 30 million tonnes)

Production is widely distributed. North America accounts for about 50 percent of the total, with Europe and Japan contributing major portions. In the Nordic and North American regions, Brazil, Chile and New Zealand, unbleached sulphate pulp predominates, while in Western Europe and other parts of Latin America and Asia the predominant component is recycled waste paper. In Australia, where waste paper based linerboard have predominated, increased quantities of kraft linerboard are now produced following increases in pine kraft pulping capacity.

The trend is toward increased use of waste paper and diminished use of unbleached sulphate pulp.

Fluting medium (production level 19 million tonnes)

In the Nordic countries, the furnish is largely semi-chemial pulp, while in North America and Eastern Europe this accounts for about 550 kg/ton of the furnish. In Western Europe, Latin America and Asia the major component is waste paper, accounting for 850 kg. Non-wood fibre may be a significant component in developing countries of Africa and Asia.

The trend will be toward increased use of waste paper and less use of semi-chemical pulp.

Folding boxboard (production level 18 million tonnes)

In Nordic countries, the furnish composition includes 200 kg of mechanical and thermomechanical pulp, 700 kg of bleached and unbleached chemical pulp, while in North America, Europe and Latin America there are 500-600 kg of recycled waste paper and little mechanical pulp per ton of production. A substantial proportion is coated, with total pigment and filler usage of 60-140 kg per ton.

The trend is toward increased use of thermomechanical pulp and bleached hardwood sulphate pulp with some decrease for mechanical and coniferous sulphate pulps.

Wrapping papers (production level 28 million tonnes)

This category is composed of roughly equal parts of kraft wrapping and other wrapping, and there are extreme variations of furnish. The unbleached sul-

Table 44 - FURNISH PROJECTIONS FOR 1995 BY IWP (million metric tonnes)

Region	Paper production	Total furnish	Ratio*	Pulp Mechanical	Chemical	Waste paper	Filler pigment	Other
Nordic	19	20.5	107	7.5	9.8	1.3	1.9	
Other West Europe	43	46.5	110	5.7	19.8	16.4	4.6	
Total West Europe	62	67	109	13.2	29.6	17.7	6.5	
North America	97	103	106	19.5	57.7	21.5	4.3	
Japan*	27	28.5	106	3.4	10.6	12.9	1.6	
Oceania	3	3.2	107	0.9	1.4	0.8	0.1	
Latin America	10	10.8	108	0.8	4.0	5.0	0.4	0.6
East Europe and USSR	17	18.7	111	2.3	10.9	4.6	0.9	
Total	216	231.2	107	40.1	114.2	62.5	13.8	0.6
Ratio*		107		18	·52	30	6	
CHASE SCENARAIO								
Total	245.9	275.9	111	32.9	130.3	83.3	(15.7)**	13.7
Ratio		111		13	53	34	(6)	6

* Kg furnish per 100 kg paper
** ( ) Kg filler and pigments as IWP projection

phate pulp content ranges from a high level of 750 kg/ton in the production of Nordic countries to 200-600 kg in other regions. Recycled waste paper ranges from 250-400 kg. Other chemical pulp makes up 100-400 kg of the furnish.

There are expected to be small declines of unbleached sulphate and possible increased use of waste paper in Nordic production.

Fluff pulp and dissolving pulp (production level 7 million tonnes)

The world volume of fluff pulp is about 2.7 million tonnes projected to increase to 4.2 million tonnes by 1995. Most of this (80-85 percent) is made from bleached sulphate, with a small proportion of hardwood. There is currently growth of chemi-thermomechanical pulp for fluff pulp.

Dissolving pulp in 1984 amounted to 5.6 million tonnes. Of this 60 percent was sulphite pulp, 30 percent sulphate. The total volume is expected to drop to about 5 million tonnes by 1995.

TABLE 44 shows an alternative projection of furnish composition based on the IWP review of developments applied to the CHASE projection of total paper production. This projection of furnish takes account of the projected production of *newsprint, printing and writing paper* and *other paper and paperboard*. For *developed market economies*, the projected composition of *printing and writing*

paper and of *other paper and paperboard* recognizes prospective changes in the furnish composition of the detailed grades. Where no projection of detailed development by grades was possible, then the approximate composition in 1984 indicated by the *FAO Pulp and Paper Capacity Survey* is retained. For certain important groups of countries, only the furnish composition of total paper production is available — notably for Japan and for Eastern Europe and the USSR. The projection for these countries is made in relation to the projected production of total paper.

Compared with the statistical projection, the IWP's estimates of furnish composition suggest generally higher levels of mechanical pulp, compensated for by lower levels of chemical pulp and waste paper. There is also a larger component of other fibre pulp in the statistical projection, largely because the IWP composition is not estimated for the countries mainly utilising other fibre pulps (FIGURES 37 and 38).

A particular feature of the IWP estimates is the stress on expansion of thermomechanical and chemi-thermomechanical pulps replacing mechanical and chemical grades.

There are two further considerations. In other (non-wood) pulp, baggasse may be less available in the future as cane sugar production may be expected to diminish. The second related to waste paper: the increasing use of additives, special coatings, glues, the introduction of laminates of other

Figure 37

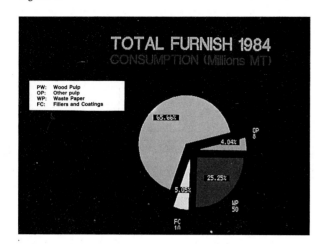

Figure 38

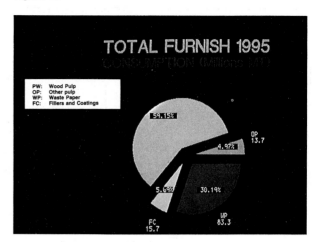

materials and developments of printing may make an increasing proportion of waste paper unsuitable for recycling. On the other hand, improved tech-nology in waste recovery such as disk separation may increase the potential for recycling.

CHAPTER 6

BALANCE BETWEEN SUPPLY AND DEMAND

In this chapter the projections of consumption and production are brought together (TABLE 45). In economic theory, suppy and demand are in an interactive relationship such that the volume delivered by producers equals the volume taken by consumers with a certain set of prices. The projections in this study were all prepared on the assumption of constant price. In theory, if projected consumption is higher than projected supply for a given year, this would result in a movement along the demand and supply curves to a point of intersection, adjusting the volume of consumption, production and the price.

In practice these projections of consumption and production on the assumption of constant prices were fairly close. As described in Chapter 7, the equality of projected supply and demand was achieved by adjusting the total supply at the world level to equal projected demand.

These projections of world consumption and production of *paper and paperboard*, assuming constant real prices indicate an increase from 187 million tonnes in 1984 to 1995 levels of 254.8 million tonnes under the *FAO scenario* and 245.9 million tonnes under the *CHASE scenario*. An experiment in simul-

taneous modelling of supply and demand at the world level projects an equilibrium volume of consumption and production in 1995 of 256 million tonnes. The Industry Working Party forecasts alternative 1995 levels of 240 to 250 million tonnes. The increases in production projected up to 1990 appear to be generally compatible with estimates of capacity expansion reported in the *FAO Pulp and Paper Capacities*.

Considering the balance in particular regions in 1984, production in the *developed market economies* exceeded consumption by about 6 million tonnes supporting a net trade flow of 5 million tonnes to *developing market economies* and a small net import by *centrally planned economies* (FIGURES 39 to 42).

Projections indicate a potential to increase the positive balance in *developed market economies*. In the *developing market* and *centrally planned economies*, the alternative scenarios relating to fixed capital formation indicate production increases which lead to a decrease in the degree of self sufficiency.

Developed countries

Developed market economies

The projected growth of production is somewhat faster than the growth of consumption. This reflects

Table 45 - **PROJECTED CONSUMPTION AND PRODUCTION OF TOTAL PAPER AND PAPERBOARD TO 1995** (million metric tonnes)

Type of Economy	Actual 1984	FAO 1995	CHASE 1995
Developed market			
Production	150.2	196.4	190.6
Consumption	144.4	177.1	172.7
Developing market			
Production	15.3	22.1	21.3
Consumption	20.1	37.0	34.7
Centrally planned			
Production	22.0	36.2	33.8
Consumption	22.8	40.6	38.5
World			
Production	187.6	254.8	245.9
Consumption	187.3	254.8	245.9

Figure 39

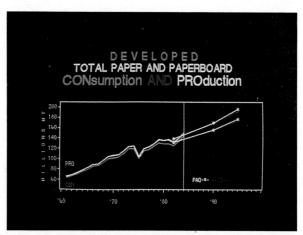

Figure 40

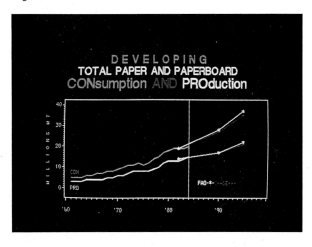

Figure 41

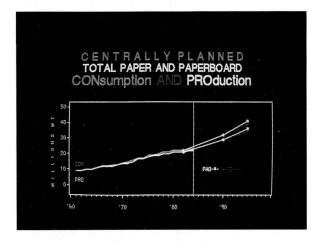

Figure 42

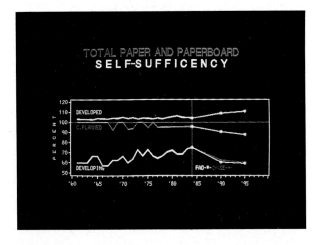

forecasts in the economic scenarios of the development of fixed capital formation.

The differences in growth of production and consumption affect the ratio of production to consumption which is defined here as self-sufficiency.

Thus, the region's self-sufficiency ratio is seen to rise in the projection to 111 percent in 1995 compared with 104 percent in 1984. The growth differences yield higher ratios in 1995 for Western Europe (112 compared with 107) and Japan (106 compared with 99). For North America the self-sufficiency ratio also rises substantially (113-115 compared with 104).

Although the continuation of past allocation of capital resources would lead to an expectation of these levels of production development, neither the market development nor the constrained forest resource situation (particularly of Western Europe and Japan), nor the high relative costs would lead one to expect a sharp upswing in net exports. Indeed it may be reasonable to expect that the self-sufficiency ratio might decline or increase only slightly in the circumstances that net import demand from other regions was to expand.

Eastern Europe and USSR

There is a slight declining tendency in self-sufficiency in the projections for Eastern Europe and the USSR. In aggregate, the resources clearly could permit maintenance of self sufficiency in these countries.

Developing countries

Africa

Paper and paperboard consumption in developing countries of Africa in 1984 was 1.3 million tonnes with average per caput assumption of 2 kg. Growth in consumption during the 1970s averaged around 4 percent with somewhat faster growth — around 6 percent — in the North African countries.

Domestic production within the region in 1984 was 0.5 million tonnes with 0.8 million tonnes of imports. All the countries of the region were net importers of paper and two-thirds were entirely dependent on imports for their supplies. In many of these countries, total consumption was as low as a few thousand tonnes. Fourteen countries have some domestic production, but in 1984 only Egypt exceeded 100 000 tonnes. Algeria, Kenya, Morocco and Zimbabwe produced 60-90 000 tonnes. The estimated installed capacity in 1984 was 1 million tonnes, giving apparent average utilisation of capacity around 50 percent.

Wood pulp production was about 300 000 tonnes of which half was the market pulp produced for export by Swaziland. Production of other fibre pulp was 50 000 tonnes. About 150 000 tonnes of pulp was imported, the remaining fibre furnish being made up of 160 000 tonnes of waste paper. The estimate of installed pulp capacity was 1 million tonnes, so that the apparent utilisation of capacity was 35 percent. The utilisation rate was low for both pulp and paper capacity. Significant installed capacity was out of commission in two countries as of 1986, and announced capacity has been subjected to protracted delays in commissioning in several other countries.

Consumption is projected to increase to 1.5 to 1.4 million tonnes by 1995, with production increasing to 530 000 million tonnes. Thus projected self-sufficiency would not increase from the current level of 38 percent. The estimate of capacity to 1990 of 1.1 million tonnes indicates that the 1990 projected level of paper production of 600 000 could be met (apparent capacity utilisation would average 50 percent).

Projected paper production is below projected consumption in the countries which have significant production capacity, thus the more efficient utilisation of established and planned capacity would be supported by the size of the domestic market. The total African market provides a potential for considerable further improvement in the utilisation of capacity already established and planned in the region. This would imply a higher level of production than projected. As the market in any one country in Africa tends to be small, the efficiency of production would be assisted by the development of exports to neighbouring countries.

The existing capacity has involved an outlay of capital resources estimated at US$1-2 thousand million. The region's annual expenditure on imports of paper in 1984 was running at some US$600 million per year. A 10 percent improvement in utilisation of existing capacity would produce paper worth around US$60 million per year. Improving utilisation to a reasonable level of 80 percent would add US$120 million/year to the industry's contribution to the region. Failure to utilise this expensive capital asset would impose a severe and unrequited economic burden in debt servicing and physical deterioration of the installed equipment. Full mobilisation of existing capacity and the quick commissioning and efficient utilisation of new capacity are matters of the highest priority. Full utilisation of estimated 1990 capacity could make the region 50 percent self-sufficient in supplying projected consumption and would save US$300 million annually in expenditure on imports. Effectively operated, the Industry is capable of making a substantial positive contribution to the economies of Africa.

Asia

Consumption of paper and paperboard in developing countries of Asia including the Near East, the Far East and centrally planned Asia totalled 16.5 million tonnes in 1984. The per caput consumption of

6 kg. During the 1970s production increased at 10 percent per year in *developing maket economies* and 7 percent per year in the *centrally planned countries*. Consumption growth for the whole region averaged 7-8 percent.

Total production in 1984 was 13.2 million tonnes so that the region was 80 percent self-sufficient. Of the 35 countries with appreciable paper consumption, 23 have some domestic production. China with 7 million tonnes and Republic of Korea and India with around 2 million tonnes are the largest producers. There are six other countries with production exceeding 100 000 tonnes, nine intermediate producers and five smaller producers with less than 10 000 tonnes/year. Total capacity in the region is 15 million tonnes and apparent capacity utilisation is rather high at 88 percent.

Wood pulp production in 1984 was 3 million tonnes and production of other fibre pulp 6 million tonnes. Fibre furnish was supplemented by importing 1.6 million tonnes of pulp and using some 4 million tonnes of waste paper. Apparent fibre consumption per ton of paper was 1.1 ton. Wood pulp capacity was 3.1 million tonnes and capacity for other fibre pulp was 6.0 million tonnes, apparent capacity utilisation was 99 percent.

Consumption is projected to increase to about 33 to 37 million tonnes by 1995 while the production level is projected to be in the range of 22.6 to 26.6 million tonnes. The higher level is associated with high rates of capital formation for Asian market economies, forecast under the *CHASE scenario*. Under these projections the region would increase its dependence on imports. Production projected to 1990 under these two scenarios — 17 and 20 million tonnes — is well within capacity estimated for the region for 1989 of 21.4 million tonnes.

The industry in this region is growing rapidly and the aspiration is to a greatly increased consumption. In the most populated and the high consumption areas — China, East Asia and South Asia — the fibre raw material supply poses a major problem for the expansion of production and the intense competition for raw material results in high relative cost of domestic raw material. Though there are certain countries in the region with relatively low levels of capacity utilisation, the major problems are to generate resources for capital renewal and expansion. In several countries, development of the industry must be accompanied by action to control environmental conflicts in intensely populated areas. In all countries, expansion must be accompanied by investment in education and training in all aspects of the industry. As well, technical, marketing and research support has to be developed. In many smaller producing countries, local development should be accompanied by linkages to supporting information systems in larger producing countries to ensure access to material that would otherwise be beyond their reasonable capability.

Latin America

Latin America in 1984 was 9.7 million tonnes with average per caput consumption of 24 kg. In the 1970s, consumption increased at a rate of 5.9 percent per year.

Domestic production within the region was 8.6 million tonnes in 1984 with imports of 2 million tonnes and exports of 2 million tonnes (nearly 90 percent self-sufficiency). Practically all of the exports are supplied by three countries; Brazil, Chile and Colombia (the first two are net exporters). Thirty other countries with significant consumption are net importers. Nineteen countries have significant production. Brazil, Mexico and Argentina have production exceeding 1 million tonnes: Venezuela, Chile and Colombia have production in the range of 400-700 000 tonnes. Regional industry capacity in 1984 was 11 million tonnes and apparent capacity utilisation 78 percent.

Production of wood pulp was 5.5 million tonnes and other fibre pulp 800 000 tonnes. The region was a net exporter of pulp to the extent of 800 000 tonnes because of exports of Brazil's exports, which approached 1 million tonnes and those of Chile 500 000 tonnes. Waste paper consumption was 3.3 million tonnes. Thus total apparent fibre furnish was 8.8 million tonnes or 1.02 tonnes per ton of paper produced. Pulp production capacity in the region in 1984 was 7.5 million tonnes and apparent capacity utilisation 84 percent.

The 1995 consumption level is projected to be 17.1 million tonnes on the *FAO scenario* and 15.2 million tonnes in the *CHASE*, while production is projected to be 12.5 and 10.1 million tonnes in the *FAO* and *CHASE scenarios*, respectively, suggesting a decrease in self-sufficiency to 73 and 66 percent This would be a reverse of the trend since 1961 when self-sufficiency of the region has increased from 66 percent to 90 percent. Projected paper production in 1990 in the range 11.8 to 12.5 million compares with the total capacity of 12.9 million tonnes predicted for 1989.

Since 1970 consumption has grown with the economies, while production has increased more rapidly, resulting in a rise in self-sufficiency from 70 to 90 percent with significant development of net pulp export. For future development, fibre supply is not considered as a constraint for the region as a whole. Several countries in the region have demonstrated comparative advantage in wood fibre production. An increase in supplies of the main non wood fibre, bagasse, may be constrained by low sugar prices and the slow development of demand for sugar. For small countries of Central America mobilising necessary volumes of wood fibre for pulp manufacture does constitute a constraint.

A constraint on industry development is the mobilisation of capital particularly in countries with major debt servicing and balance of payment problems. The advantage is with countries which can mobilise domestic capital and have internal capacity to produce pulp and paper machinery. For many countries in the region, the size of domestic market is small in relation to the scale of an efficient operation. Trained personnel, market intelligence and research support are components requiring development either within the producing countries or by the establishment of effective linkage to support elsewhere.

Conclusions on supply and demand balance

In practice future market clearing actions by consumers and producers will ensure that the amount produced corresponds to the quantity consumed. The projected balance implies an increase in net exports by *developing market economies* and a diminution of the self-sufficiency of *developing market economies* and *centrally planned economies*.

To illustrate an alternative possible direction in the relative expansion of production and consumption, regional production is adjusted to correspond with recent trends in the development of net exports and self-sufficiency (TABLE 46). In relation to the *CHASE scenario* of consumption, *developing market economies* and *centrally planned economies* increase their production. The self-sufficiency in the

Table 46 - TOTAL PAPER AND PAPERBOARD CONSUMPTION, PRODUCTION BALANCE TO 1995 Projections compared with an alternative reflecting trend in net trade (million metric tonnes)

Type of Economy	Actual 1984	SCENARIOS		
		FAO	CHASE 1995	Alternative
Developed market				
Production	150	196	190	181
Consumption	144	177	172	172
Developing market				
Production	15	22	21	28
Consumption	20	37	34	34
Centrally planned				
Production	22	36	33	36
Consumption	23	40	38	38
World total				
Production	187	254	245	245
Consumption	187	254	245	245

developing world increases from the 1984 ratio at a rate similar to the past decade. The developed market economies maintain the ratio of 1984, while self-sufficiency of *centrally planned economies* decreases slightly again in line with the change over the past decade (FIGURES 43 to 45).

Figure 43

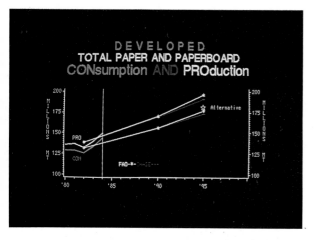

Figure 44

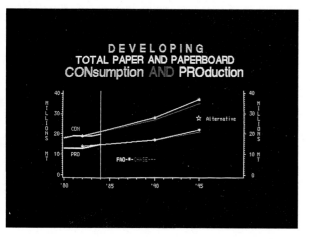

Figure 45

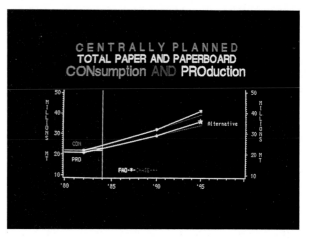

The projection of fibre furnish and wood pulp consumption adjusted in relation to constrained production, is shown in TABLE 47.

Table 47 - FIBRE FURNISH CONSUMPTION IN 1995 UNDER THE ALTERNATIVE PRODUCTION SCENARIO (million metric tonnes)

Product planned	Type of economy			
	Developed world	Developing market	Centrally market	Total world
Total fibre furnish	196	32	31	260
Mechanical pulp	29	1	3	33
Chemical pulp	102	12	16	130
Other fibre pulp	1	5	8	14
Waste paper	64	14	5	83
Pulpwood (million m^3)	582	61	87	730
Total paper and board production	181	28	36	245

CHAPTER 7

MODELLING DATA AND ASSUMPTIONS

Central to this study are the projections of production and consumption of paper and paperboard. Preparation has involved considerable effort on the development of econometric models and the improvement of the data available to support these. This chapter describes the basic principles underlying the approach, the data that has been used and the models behind the projections.

This *FAO Outlook Study* is built on the analysis of the historic relationship between consumption and production of paper products, and measures of economic performance. The analysis is carried out within the framework of economic theory relating to the demand for goods and services by consumers and their supply by producers. The strucure of econometric models was constrained by availability of consistent international data sets, both for the products and for the appropriate economic variables. Having identified relationships which are consistent with theory and statistically acceptable, these models then provide the mechanism of calculating reference projections from selected scenarios of future development of economies.

Behind the projections lies the assumption that the relation between consumption and production and the economic variable will remain the same for the future period as it was for the period of the historical analysis.

Approach to demand modelling

A two-stage approach to demand modelling was adopted with the aim of retaining a simple model structure while at the same time achieving the greatest possible definition of individual products. The first stage was to estimate aggregate models for product groups or sectors for which international data was available for all countries, relating consumption to economic indicators. On this basis, reference projections of demand were established for *newsprint, printing and writing paper*, and *other paper and paperboard*.

The second stage established demand models for detailed components of the broad product aggregates, and relate consumption of the component products to economic indicators that were more precisely identified with the sector in which those products are used. This disaggregated modelling was possible only for member countries of the OECD for which detailed data, both on products and on the economy, were available.

The examination of product detail was taken further in a mainly qualitative manner by experts from the Industry Working Party (IWP). In addition, trend analysis of consumption of the main products by countries was carried out to determine whether this could provide a stable indicator of development of product consumption without reference to economic development (Graff, 1984).

Data for the analysis

A major consideration in deciding the approach to modelling was the availability of data. Very important to this study was the improvement in quality of paper product data achieved through detailed examination of the historical data by industry experts. The data series used are described below.

Consumption, production and price

For the aggregate modelling of consumption and supply the product data of the *FAO Yearbook of Forest Products* for three product categories — *newsprint, printing and writing paper* and *other paper and paperboard* — were used. These data series contain information on production and trade volume and trade value for 146 countries in a consistent form from 1961 to 1984.

Consumption data used in the modelling is apparent consumption; that is, production plus imports minus exports. Changes in stock levels are not considered.

An indication of price is provided by the unit value of imports and exports (total value divided by the volume).

Detail on product subdivisions

For the *developed market economies* in the Organisation for Economic Co-operation and Development (OECD), fairly consistent series were available from the *OECD Pulp and Paper Statistics*. These data covered production and trade volume in coated and uncoated *printing and writing paper*, household and sanitary paper, and about 10 further sub-divisions of wrapping and packaging paper and *other paper and paperboard*. Based on these data, a demand modelling exercise was carried out on coated and uncoated *printing and writing paper*, household and sanitary, and on three broad groups of wrapping and packaging paper — linerboard and corrugating medium, folding boxboard and wrapping paper. Limitations of these detailed series were the short period (not more than 10 years) for which consistent data were available, and the absence of trade value or any other statistic indicative of price.

Newsprint basis weight

Because of its possible importance in explaining trends in total tonnage consumption of *newsprint*, a special survey was made of developments in *newsprint* basis weight (grammage) for 12 countries.

Data on past development of industry capacity and a medium term forecast were those provided by *FAO Pulp and Paper Capacities*.

Pulp and other inputs

In calculating the pulp requirements to supply projected paper consumption, estimates of the input of particular types of pulp, waste paper and other material inputs such as fillers, coatings and pigments were required. A statistical relationship between the production of the main paper grades and types of pulp, waste paper and fillers was estimated. Projections of pulp consumption were made applying these relationships to the projected production of the paper grades. The relationship of specified pulp, waste paper, and filler and pigment input to specific paper grades was reviewed by the (IWP).

Data on pulp production and apparent consumption — subdivided into the main grades of mechanical, chemical and semichemical wood pulp and pulp of other fibre — is available from the *FAO Yearbook of Forest Products*. Waste paper data has also been assembled in this series. These data, however, do not relate the total volume of pulp and waste paper consumed to particular types of paper, though some information on this relationship is provided by the *FAO Waste Paper Survey*. Two special surveys were carried out to augment information in this area. The industry in more than 20 countries answered a survey of total furnish composition and the data was reviewed by the IWP. A study of of fillers, coatings and pigments (Galsworthy, 1984) covering 28 countries and country groups, and concentrated particularly on their use in *printing and writing paper*.

Wood consumption

Assessing the relationships between industry development and the supply of wood from the forest is an important function of forest sector outlook studies. The estimate of wood required specifically for the pulp and paper industry is difficult since there are only a few cases where that industry's wood consumption is recorded separately from sawmilling or other industrial consumption.

In this study, an approximate projection of industrial wood consumption by the pulp and paper, sawnwood and wood-based panels industries has been prepared to provide a total which could be related to the wood supply situation. The data for this estimate were the production of sawnwood, wood based panels, wood pulp and consumption of industrial roundwood from the *FAO Yearbook of Forest Products*. The projections are based on the projected production of these products. Parallel projections of sawnwood and panels to those for paper have been used together with projections of consumption of pulp requirement mentioned above. This projection of industrial roundwood does not attempt to identify net trade.

Economic indicators

Gross domestic product (GDP) was adopted as the basic economic indicator available for all countries on a consistent basis. The source selected for this data was the *FAO Compendium of Macro-Economic Indicators*. It provides consistent historical series of GDP at 1980 prices for the years 1961-81 for all countries of the world. The data are derived from the *UN Compendium of World Development Indicators*, from OECD, the World Bank and the Economic Commission for Europe (ECE). As a base line for the future development of GDP, the *FAO Compendium* was updated to 1984 making use of data and short-term forecasts of OECD, United Nations Conference on Trade and Development (UNCTAD) and the World Bank. FAO did no original work on data collection but did ensure standardisation of the data series. These data are used with the reservation that they were the best available consistent time series.

The reservations of the *FAO Compendium* authors relating to imperfections of real product as an indicator of economic performance, the failure of prices used to reflect actual utility, the discontinuity of incomplete data in some series, and the varying quality and basis of data, should be noted.

GDP forecasts. The preparation of the reference projections made use of two alternative scenarios for the growth of GDP for the period 1985-95 (TABLE 48). As one alternative, the FAO adjusted trend scenario was adopted, referred to here as the FAO scenario.

As an alternative scenario of economic development (and as a check on the *FAO scenario*), a second scenario was included from an independent source regarded by the IWP as a representative commercial view of economic development. For this purpose, *Chase Econometrics' 1986 Long Term Report* was adopted. Referred to here as the *CHASE scenario*, this provides specific economic forecasts for 60 countries and an indication of a growth estimate for *centrally planned economies*. For countries not covered in this scenario, the FAO adjusted trend scenario was retained.

Detailed economic data. Historic data on detailed indicators of economic activity are contained in the *United Nations Compendium of World Development Indicators* and in OECD economic data for some *developed market economies*. Specific forecasts for some of these variables are included in the *Chase Long Term Report*. These sources were utilised in the disaggregated analysis.

Economic data for supply modelling

Three time-series of gross fixed capital formation (GFCF) were estimated for each producing and/or consuming country: an historical series, a high growth forecast and a low growth forecast. The International Monetary Fund, World Bank and United Nations Statistical Office compile estimates of GFCF and GDP in local currency. For the years 1961-1984, the proportion of GDP expended on capital formation was related to GDP from the *FAO Compendium of Macro-Economic Indicators* and a mean ratio of GFCF to GDP calculated for each country. The ratio from the historical series was multiplied by the projection of GDP from the FAO Compendium to yield a constant dollar series on GFCF to 1995. The implicit growth rates in this series are labelled the *FAO scenario*.

Chase Econometrics provided a forecast series on capital formation in constant local currency for 59 countries. In the *CHASE scenario*, the growth rates of capital formation implicit in the Chase Econometrics' outlook replace, for the countries for which they are available, the rates derived from the FAO Compendium.

The *FAO* and *CHASE scenarios*, previously defined, were adopted as alternatives. They were of course developed quite independently. Though in general it may be said that the indications of future growth in the *FAO scenario* are higher than those in the *CHASE scenario*, there are exceptions. Notable among these are the Chase GDP and GFCF forecasts for USA and Canada for the period 1982 to 1990, and for some developing countries.

The IWP reviewed the projections of GDP growth (TABLE 48). Several contributors regarded the two scenarios as acceptable alternatives representing a range of possible development. Some regard the high alternative as acceptable for their country. A respondent from the centrally planned region regards the low alternative as unrealistic for those countries, while the high was acceptable. Some others preferred the low alternative. Several respondents from European countries consider both scenarios as over-optimistic.

Table 48 - GROWTH RATES OF GROSS DOMESTIC PRODUCT TO 1995 IN SELECTED COUNTRIES (in percent per year)

Country	Actual 1970-82	FAO scenario 1982-90	FAO scenario 1990-95	CHASE scenario 1982-90	CHASE scenario 1990-95
Canada	3.5	2.7	3.4	3.1	3.0
USA	2.4	2.9	2.7	3.5	2.6
Finland	3.1	3.1	3.6	3.2	2.8
France	2.9	2.5	3.8	1.9	2.5
Germany, Fed. Rep.	2.2	2.1	2.9	2.1	1.8
Portugal	4.7	2.8	4.1	1.5	3.0
Spain	3.6	2.3	3.5	2.7	2.7
Sweden	2.0	2.1	3.3	2.1	3.3
United Kingdom	1.4	1.8	2.3	2.3	3.2
Australia	2.7	2.5	3.7	3.1	2.8
New Zealand	2.4	1.3	3.4	2.7	2.6
Japan	4.0	4.8	4.8	3.5	3.6
Argentina	1.0	3.5	2.2	2.4	3.2
Brazil	7.3	5.0	6.0	2.4	4.7
Chile	1.8	3.0	3.0	3.2	4.0
Colombia	5.1	4.3	5.4	3.2	4.0
Indonesia	7.3	4.7	5.4	5.1	5.2
Philippines	5.7	5.1	6.7	1.0	3.7
Thailand	7.7	6.1	6.0	5.7	5.7
Czechoslovakia	3.1	2.2	3.5	2.0	2.0

Demand and supply models

Modelling in this *FAO Outlook Study* is consistent with the idea that the volume of paper that people consume varies with its price, while the volume that producers will supply depends on the price and their costs of production. The market sets a price at which the volume consumed is matched by production. Over time, as people's income changes, the volume they would like to consume at a given price shifts. Producers also respond to changes in the level of economic activity. Paper is only one of many products and, as technology changes, people may change their preference for paper compared with other means of fulfilling their need for communications and packaging.

The demand model

The demand model relates consumption to price, income and time. It concentrates on the longer term and demonstrates a strong relationship between consumption and income development (represented by GDP) called income elasticity. There is also a relationship between consumption and price: price elasticity. Though significant, the price elasticity is small — demand may be described as price inelastic — so estimation of price change is not critical in projecting consumption.

The complex of changes in technology, competition from other products and changed preferences has resulted in consumption changes not closely related to income or price movements; these have been represented by a relationship to time.

The functional relationship for *newsprint, printing and writing paper* and *other paper and paperboard* is given in TABLE 49. It was found more accurate to estimate a separate income elasticity for five country groups. Four include countries in the per caput income ranges indicated, while the fifth is the United States, which was separated because of its large share in world consumption of paper.

Projections of consumption. In making projections for consumption of each product for each country, the income elasticity for the income group within which the country falls and the respective time trend is applied (TABLE 49). Price is assumed constant. Consumption is projected according to two assumptions of GDP growth for the country, namely those of the *FAO scenario* and the *CHASE scenario.* The volume in the "base year" from which future consumption is projected is the average annual volume of consumption over the five year period 1980-84, and the base year used is 1982. Regional consumption is the sum of all countries' consumption in that region.

Uncertainty. The models are, of course, imperfect representations of the relationship of consumption to income, so the projections are subject to statistical error. For each product the projection for the world has upper and lower limits of ±4 percent, while the confidence limits on projections for regions are wider — ±6 to ±15 percent. The projections of consumption in individual countries for individual years are in a much wider range. For large consumers such as the USA, this range is ±8 to 12 percent, while for smaller consumers ±25 percent is more characteristic.

The uncertainty is greatest for projections of any single year's consumption for an individual country and progressively less where it relates to the trend over a period of years for a region. The projection estimates are, of course, also conditional on the estimated growth rates of GDP, which fact tends to increase projection uncertainty.

The economic forecasts are also subject to uncertainty, and these projections should be regarded as indications of a growth path within a range of uncertainty.

Disaggregated demand analysis. Within the broad aggregates of *printing and writing paper* and *other paper and paperboard* there are components with very different technical characteristics, which relate to different end-uses and may therefore be better identified to particular areas of economic activity than to the overall indicator GDP. Some further analysis was carried out of consumption in *developed market economies* using *OECD Pulp and Paper Statistics.*

The categories examined were:
— Coated printing and writing papers
— Uncoated printing and writing papers
— Household and sanitary papers
— Containerboard (liner and fluting)
— Folding boxboard
— Wrapping

Making use of detailed economic indicators, *printing and writing paper* and household and sani-

Table 49 - INCOME ELASTICITIES AND TREND FACTOR USED TO PROJECT FUTURE CONSUMPTION

Grouping		Income elasticities		
		Newsprint	Printing and writing	Other paper and paperboard
Lowest income	Group 1	.83	.83	1.27
	Group 2	.68	.91	.75
	Group 3	.60	.70	1.33
Highest income	Group 4	.59	.74	1.27
	USA	.43	.23	1.13
Trend factor: yearly change in consumption (percent)		.6	1.8	-1.3

Equation used to calculate income elasticity
and yearly change in consumption:

$$\begin{array}{l} \text{(income-elasticity)} \\ \text{Consumption} = \text{Intercepts x GDP} \\ \qquad\qquad\qquad \text{(price elasticity)} \\ \qquad\qquad\quad \text{x price} \\ \qquad\qquad\qquad \text{(yearly change in consumption)} \\ \qquad\qquad\quad \text{x year} \\ \qquad C = a \ x \ GDP \\ \qquad\qquad x \ P^c \\ \qquad\qquad x \ t^d \end{array}$$

tary paper are related to GDP, folding boxboard to private consumption and containerboard and wrapping paper are related to industrial production. Absence of data on prices and the short time series prevented consideration of either a relationship to price or the inclusion of a time trend in the analysis.

Further commentary. The selection of appropriate economic indicators is an important consideration. GDP provides a broad indication of general development of the economy to which all aspects of economic activity contribute and relate.

Paper consumption relates to many aspects of the economy — to communications, private consumption expenditure, industrial production and to public and private sector service activities. There may be appreciable differences between development of particular areas of the economy, both in particular periods of years and in general over time. Thus, industrial production may fluctuate more sharply between slump and boom than the sum of all economic activity. Service activity in the economy may tend to increase its share of the total economy over time. Because of the limit on availability of detailed indicators of economic development, when one considers all countries of the world and the absence of estimates of future development of other economic indicators, GDP has been adopted as the economic variable in the aggregate analysis. Some exploration of other indicators has been attempted in the disaggregated analysis of some products for *developed market economies* and in the work of the IWP.

The price of paper varies between countries and over time. Over the period from 1960 to 1984 there was a general tendency to a decrease in the real price of paper, but with appreciable fluctuation resulting in peaks in 1974 and 1981 and troughs in 1972, 1979 and 1983. There is a significant variation in the price between countries depending on whether consumption derives mainly from domestic production or from imports, on the cost of production, cost of delivery and special conditions such as tariffs and price regulation. In constructing the model, import unit value has been taken as an indicator of domestic prices where imports dominate domestic supply, and export unit value where local production is dominant. This may underestimate the prices actually determining domestic consumption when the market is highly protected and may tend to overestimate domestic price in countries with large domestic production as trade may be concentrated in high value grades.

Certain aspects of the change in consumption over time and the difference in consumption between countries could not be explained by price and income. For example, prevailing tendencies in paper consumption which may result from changes in information, packaging or transportation, technology and the development of competing materials were represented in the model by the relationship to time.

The explanation of variation in the level of consumption between countries was greatly improved by the introduction of country specific variables.

In many countries the growth in consumption of paper products was lower in the later years than in the earlier years of the period examined in this study. It is also observable that, taking that period, the ratio of consumption to income is lower in later years for some products. In economic terms this means that the income elasticity is less than 1.0. It has also been suggested that the change in consumption per unit change in income is smaller in later than in earlier years. This would imply a decline in income elasticity over time. The model adopted includes, as mentioned above, a simultaneous relationship to income, to price and a time trend. It appears that the crude time trend and the shift in consumption income ratio, mentioned above, are accounted for by changes in income growth, price and a more narrowly defined time trend. The time trend in the integrated model may be related to technological change, and to trends in end use and substitution which are not particularly related to price or income development. With careful statistical analysis to examine the point, it was not possible to find a significant change in the income elasticity over the 23-year period in the fully specified model.

The supply model

The supply model relates production to gross fixed capital formation (GFCF), determining an output elasticity, and the price of inputs, represented by the price of pulps, determining a price elasticity. Nations are treated as producers, and in constructing the model countries have been grouped according to the degree of dependence on domestic production to meet consumption needs, separating net exporters. The available international data does not provide for an adequate representation of the comparative advantage of different producers. This model is put forward as a tentative first step in international modelling of supply.

Projections of production

In making projections of production of each product for each country, the output elasticity to GFCF for the group in which the country falls is applied (TABLE 50). Price is assumed constant. Production is projected according to two assumptions of growth in capital formation, namely the *FAO* and *CHASE scenarios.* The volume in the "base year" from which future production is projected is the average annual volume of production over the five-year period 1980-84 and the base year is 1982. Regional production is the sum of all countries' production within that region.

Table 50 - OUTPUT ELASTICITIES USED IN THE PROJECTION OF SUPPLY (for changes in Gross Fixed Capital Formation)

Production Consumption Ratio		Newsprint	Printing and writing	Other paper and paperboard
Low	Group 1	0.92	1.08	1.17
	Group 2	0.85	0.99	0.86
	Group 3	0.78	1.01	0.72
High	Group 4	0.42	1.04	0.88
	Canada	0.31	1.04	0.88
	USA	0.92	1.21	0.46
	Finland	0.52	1.26	0.59
	Sweden	0.31	1.26	0.88

Equation used to calculate output elasticity for each product group:

$$\text{Production} = \text{Intercepts} \times \text{GFCF}^{(\text{output elasticity})} \times \text{pulp price}^{(\text{price elasticity})}$$

$$P = a \times GFCF^b \times P^c$$

Further commentary on supply model

The modelling of supply builds on the premises that the quantity of paper offered by producers reflects the size and profitability of domestic and foreign markets. In a given year, the comparative advantage or disadvantage of each producer is a function of its labour and raw material costs. Over time, the producer expands or contracts output based on the outlook for individual products, on the cost of capital for maintenance and expansion and on the return on alternative investments.

National production is taken from the *FAO Yearbook of Forest Products*. With the exception of the few countries which produce mainly for export, most nations expand paper production based on their domestic requirements. Critical variables which influence the level of national production are the relative level and rate of formation of social and private infrastructure. Capital formation is significantly correlated with paper production, and partially explains the difference in paper output among countries and the increase in production within countries over time. This relationship reflects the fact that embodied infrastructure is less mobile than labour, entrepreneurial talent or raw material.

For the paper producing nations which trade much of their output, foreign demand for paper and their comparative advantage or disadvantage also plays a role in production decisions. The trading nation's competitiveness depends on exchange rates, transport costs, trading preferences and macro-economic policy, to name only a few important variables. Comparable data on the cost of paper manufacture were not available for all producing nations over the period 1961-84, therefore adequate measures of comparative advantage could not be modelled.

Pulp price is used in the model as a proxy for comparative advantage in raw material costs. Product price was not included in the specification since the trade unit values for pulp and the paper products are highly correlated.

Many things could make the proportional response of output to increases in the aggregate level of capital formation vary among nations. For example, the scale and maturity of large exporters might make them dependent more on world economic growth than on changes in their domestic economies. On the other hand, producers who primarily satisfy domestic consumption react to growth in their own economies. Some economies may give priority to expanding paper production in their investment plans. They may, for example, have policies to increase literacy and consequently emphasize increased investment in paper production to support such programmes.

Supply-demand equilibrium

While the individual equations perform well, their integration is less than the theoretical optimum. There is neither interplay between supply and demand nor interaction among products. The supply projections can outpace the consumption estimates and, at the country level, countries without any apparent comparative advantage can increase their market share beyond reasonable expectations. To correct these deficiencies, models of world demand and supply were constructed for *total paper and paperboard*, *newsprint* and *printing and writing paper*. The coefficients are summarized in TABLE 51.

58

Table 51 - ESTIMATED COEFFICIENTS FROM THE WORLD EQUILIBRIUM MODEL

	Total paper and paperboard	Newsprint	Printing and writing paper
Demand			
GDP elasticity	.96	.72	1.18
Paper price elasticity	-.39	-.08	-.45
R^2	.99	.97	.97
Supply			
Capital formation elasticity	.78	.71	1.29
Paper price elasticity	.91	.57	6.36
Pulp price elasticity	-.60	-.34	-2.72
R^2	.97	.94	.63

These combined models yield several important conclusions:

1. The elasticities estimated in the individual equations of supply and demand, and used to project country output, are within range of those emerging in a combined model.

2. The price elasticity of supply is several times greater than the price elasticity of demand. This implies that most of the adjustment of the market seeking equilibrium occurs on the supply side. Comparatively, the consumption of paper is more fixed by GDP and less sensitive to price than is supply which is quite responsive to price movements.

3. In the simulations, the markets clear with little change in real product price. This implies that given the forecasts of GDP and capital formation, it is acceptable to assume constant real prices in the projections.

Generally, the projections of future production development are made on the basis of the elasticities of gross fixed capital formation related to the FAO and CHASE scenarios of economic development. In some cases, the CHASE scenario forecasts a rate of capital formation which greatly exceeds the country's rate of growth in GDP. When a country becomes a net exporter of paper, infrastructure as measured through capital formation is no longer a major constraint to increasing production. Instead, to net exporters, growth in the world market controls the development of production. To reflect this phenomenon, growth in domestic GDP replaced growth in capital formation in calculating the projections for net exporters.

After establishing the country projections for production, a world total is found for each projection year. The country projections are adjusted up or down by the ratio of world consumption to production for that year. This action derives from the finding in the world model that most of the market adjustment takes place on the supply side. In general, the adjustment is small because the two projections are usually within 5 percent of each other.

ANNEXES

ANNEX I

List of Countries Included in the Economic Classes and Regions

DEVELOPED MARKET ECONOMIES

NORTH AMERICA: Canada, United States

WESTERN EUROPE: Austria, Belgium, Denmark, Finland, France, Germany (Federal Republic), Greece, Iceland, Ireland, Italy, Luxembourg, Malta, Netherlands, Norway, Portugal, Spain, Sweden, Switzerland, United Kingdom, Yugoslavia

OCEANIA: Australia, New Zealand

OTHER DEVELOPED MARKET ECONOMIES: Israel, Japan, South Africa

DEVELOPING MARKET ECONOMIES

AFRICA: Algeria, Angola, Benin, Botswana, Burkina Faso, Burundi, Cameroon, Cape Verde, Central African Republic, Chad, Congo, Côte d'Ivoire, Djibouti, Equatorial Guinea, Ethiopia, Gabon, Gambia, Ghana, Guinea-Bissau, Guinea, Kenya, Lesotho, Liberia, Madagascar, Malawi, Mali, Mauritania, Mauritius, Morocco, Mozambique, Niger, Nigeria, Reunion, Rwanda, Sao Tome and Principe, Senegal, Sierra Leone, Somalia, Swaziland, Togo, Tunisia, Uganda, United Republic of Tanzania, Zaire, Zambia, Zimbabwe

LATIN AMERICA: Argentina, Bahamas, Barbados, Belize, Bolivia, Brazil, Chile, Colombia, Costa Rica, Cuba, Dominica, Dominican Republic, Ecuador, El Salvador, French Guiana, Guadeloupe, Guatemala, Guyana, Haiti, Honduras, Jamaica, Martinique, Mexico, Netherlands Antilles, Nicaragua, Panama, Paraguay, Peru, Suriname, Trinidad and Tobago, Uruguay, Venezuela

NEAR EAST:

AFRICA: Egypt, Libya, Sudan

ASIA: Afghanistan, Bahrain, Cyprus, Democratic Yemen, Iran, Iraq, Jordan, Kuwait, Lebanon, Qatar, Saudi Arabia, Syria, Turkey

FAR EAST: Bangladesh, Bhutan, Brunei Darussalam, Burma, Hong Kong, India, Indonesia, Laos, Macao, Malaysia, Nepal, Oman, Pakistan, Philippines, Republic of Korea, Singapore, Sri Lanka, Thailand

OTHER DEVELOPING MARKET ECONOMIES: Fiji, French Polynesia, New Caledonia, Papua New Guinea, Samoa, Solomon Islands, Tonga, Vanuatu

CENTRALLY PLANNED ECONOMIES

ASIA: China, Democratic Kampuchea, Democratic People's Republic of Korea, Mongolia, Viet Nam

EASTERN EUROPE AND USSR: Albania, Bulgaria, Czechoslovakia, German Democratic Republic, Hungary, Poland, Romania, USSR

ANNEX II

List of Participants in the Study

FAO Advisory Committee of Experts on Pulp and Paper

Hans Heidkamp
Celulosa Argentina SA
Argentina

Neil Shaw
AMCOR LIMITED
Australia

Horacio Cherkassky
Associacao Nacional dos Fabricantes de Papel e
   Celulose
Brazil

Joseph Nako
CELLUCAM
Cameroun

Howard Hart
Canadian Pulp and Paper Association
Canada

Hu Shouzu
Bureau of Foreign Affairs
Ministry of Light Industry
China

Gustavo Gomez
Carton de Colombia
Colombia

Vaclav Kubelka
Pulp and Paper Research Institute
Czechoslovakia

Peter Dauscha
Zanders Feinpapiere AG
Federal Republic of Germany

Claes von Ungern-Sternberg
Central Association of Finnish Forest Industries
Finland

Alain Arnaud
La Cellulose du Pin
France

N.S. Sadawarte
Central Pulp Mills Ltd.
India

Tokuo Hashimoto
Japan Paper Association
Japan

Warren Hunt
New Zealand Forest Products Ltd.
New Zealand

Pedro Picornell
PICOP
Philippines

Julio Molleda
Empresa Nacional de Celulosas SA
Spain

Bo Wergens
Swedish Pulp and Paper Association
Sweden

Martin Grose
Wiggins Teape Overseas Ltd.
UK

Louis Laun
American Paper Institute
USA

Youssuf Fouad
International Finance Corporation
USA

Subgroup of the committee

Marcello Pilar
CIBRAP
Brazil

Industry Working Party

Ian Chenoweth
Canadian Pulp and Paper Association
Canada

Ernesto Ayala
Compania Manufacturera de Papelas y Cartones
    SA
Chile

Timo Teras
FINNCELL
Finland

Torsten Nykopp
FINNCELL
Finland

Oskar Haus
VDP Germany
Germany FR

A. Soetikno
Indonesia Pulp and Paper Association
Indonesia

Rui Ribeiro
PORTUCEL
Portugal

Bernt Stenberg
Swedish Pulp and Paper Association
Sweden

Goran Wolfahrt
Swedish Pulp and Paper Association
Sweden

Uthen Phisuthiphorn
The Siam Kraft Paper Co. Ltd.
Thailand

E.D. Peacock
The British Paper and Board Industry Federation
UK

David W. Thornton
Wiggins Teape
UK

Irene W. Meister
American Paper Institute
USA

Anibal Valero Diaz
APROPACA
Venezuela

Gustavo J. Larrazabal
CICEPLA
Venezuela

Jorge Alvarez Gallesio
Celulosa Argentina SA
Argentina

Agustin Viale
Celulosa Argentina
Argentina

Roberto Iglesias
Asociacion de Fabricantes de Celulosa y Papel
Argentina

Roberto Jose Ranieri
Witcel S.A.C.I.F.I.A.
Argentina

Barry La Fontaine
The Pulp and Paper Manufacturers Federation of
    Australia Ltd.
Australia

Judith Maxwell
Bureau of Agricultural Economics
Australia

Peter Bennett
Australian Paper Manufacturers Ltd.
Australia

Stephen Parsons
Bureau of Agricultural Economics
Australia

Kay Kidd
Associated Pulp and Paper Mills Ltd.
Australia

Brian Gibson
Australian Newsprint Mills Ltd.
Australia

Wayne Latham
Australian Newsprint Mills Ltd.
Australia

Neil Hunter
Australian Paper Manufacturers Ltd.
Australia

Mauro A. Cerchiari
Champion Papel e Celulose
Brazil

Jose Carlos Bim Rossi
Assoc. Nacional Papel e Celulose
Brazil

Belmiro Ribeiro Da Silva Neto
Assoc. Nacional das Fabricantes de Papel e
  Celulose
Brazil

Newton Serebrenick
Aracruz Celulose S.A.
Brazil

Aliesio Grasso da Costa
Aracruz Celulose S.A.
Brazil

Raul Calfat
Industrias de Papel Simao S/A
Brazil

Donald Ross Da Mota
Industrias Klabin de Papel e Cellulose SA
Brazil

Paulo M. do Valle
Companhia Ind. de Papel Pirahy S/A
Brazil

Thomaz Lowenthal
PISA - Papel de Imprensa S/A
Brazil

Ricardo do Valle Dellape
PISA - Papel de Imprensa S/A
Brazil

Paulo T. Ribeiro
SPP NEMO - Comercial Exportadora (Suzano's
  Group)
Brazil

Murilo Ribeiro Araujo
Cia, Melhoramentos de Sao Paulo
Brazil

V.I. Suchek
Jaakko Poyry Engenharia Ltd.
Brazil

C.V. dos Santos Filho
PISA - Papel de Imprensa S31
Brazil

J.C. Rossi
Associacao Paulista dos Fabricantes de Papel e
  Celulose
Brazil

Don Gilmore
Domtar Incorporated
Canada

David Wilson
Canadian Pulp and Paper Association
Canada

Dewar Cooke
Macmillan Bloedel Ltd.
Canada

Guillermo Mullins
Compania Manufacturera de Papeles y Cartones
  SA
UK

Hou Yanzhaou (Ms)
Ministry of Light Industry
China

Renato Adriasola
Industrias Forestales S.A.
Chile

Oriana Caceres Perea (Ms)
Corporación Chilena de la Madera
Chile

Sergio Colvin
Compañía Manufacturera de Papeles y Cartones
  S.A.
Chile

Ignacio Del Rio
Industrias Forestales S.A.
Chile

Gonzalo Fernandez
Industrias Forestales S.A.
Chile

Jorge Garnham
Celulosa Arauco y Constitucion S.A.
Chile

Alejandro Hartwig
Industrias Forestales S.A.
Chile

Mario Ruz
Compania Manufacturera de Papeles y Cartones
  S.A.
Chile

Antonio Tuset
Celulosa Arauco y Constitucion S.A.
Chile

Andre Van Bavel
Celulosa Arauco y Constitucion S.A.
Chile

Guillermo Mullins
Compania Manufacturera de Papeles y Cartones
  SA
UK

Gilberto Benitez
Propal S.A.
Colombia

Victor Giraldo
Pulp & Paper Division
Colombia

Jesus Alberto Guevara
Carton de Colombia S.A.
Colombia

G. Moreno
Cartonde Colombia
Colombia

Karl Olof Kosk
Finnboard
Finland

M. Malmipohja (Ms)
FINNPAP
Finland

Antti Rytkonen
Finnish Forest Industries Central Association
Finland

L. Parikka
Central Association of Finnish Forest Industries
Finland

Mirja Soderstrom (Mrs)
FINNPAP
Finland

Bernard Majani
AUSSEDAT-REY
France

Peter Graff
Feldmuehle AG
Germany FR

Herbert von Loebel
VDP Germany
Germany FR

Ichiro Ushiba-
Dai Ei Papers Ltd.
Germany FR

Kaha Haryopuspito
APPIC
Indonesia

M. Taniguchi
Japan Pulp and Paper Co. Ltd.
Japan

Akira Takahashi
Japan Paper Association
Japan

Katsuhisa Yamada
Oji Paper Co. Ltd.
Japan

Hiromi Koshikawa
Japan Pulp and Paper Corporation (USA)

Masao Taniguchi
Japan Pulp and Paper Corporation (USA)

Kiyoshi Nagai
Jujo Paper Co. Ltd.
Japan

Takashi Mizutani
Chuetsu Pulp Industry Co. Ltd.
Japan

Hiroshi Nakamura
Honshu Paper Co. Ltd.
Japan

Kenji Okawa
Japan Paper Association
Japan

Akio Kimura
Japan Paper Association
Japan

Robert T. Fenton
New Zealand Embassy
Japan

S. Nagakawa
Wiggins Teape Research Development Ltd.
Japan

Chung Il-Hong
Korea Paper Manufacturers Association
Korea Republic

H. Ismail
Heavy Industries Corporation
Malaysia

C. McKenzie
New Zealand Forest Products Ltd.
New Zealand

Graeme N. Scales
Tasman Pulp and Paper Co. Ltd.
New Zealand

Sergio Milion
Sociedad Paramonga Ltda. S.A.
Peru

Pedrito M. Aragon
PICOP Trading Corporation
Philippines

P. Malm
Holmens Bruk AB
Sweden

B. Allguren
Billerud Uddeholm AB
Sweden

Lars Lindgren
MoDoPapper AB
Sweden

Jan Carlstrom
Scanpapp
Sweden

Karl Fredrik Karlsson
Scanfin
Sweden

Lars Thornander
Scanpapp
Sweden

Leif Karlsson
Swedish Pulp and Paper Association
Sweden

Bengt Holmberg
AB Statens Skogsindustrier
Sweden

Per Jerkeman
Celpap AB
Sweden

Otto Silfverberg
Swedish Pulp and Paper Association
Sweden

C.J. Bergendahl
Heimdalsvaegen 12
Sweden

Ola Virin
Federation of Swedish Industries
Sweden

Marianne Svensen (Ms)
Swedish Pulp and Paper Association
Sweden

Halfdan W. Mathiesen
Scansulfit
Sweden

C. Prins
ECE/FAO Joint Agriculture and Timber Division
Switzerland

Uthen Phisuthiphorn
The Siam Kraft Paper Co. Ltd.
Thailand

Derrick E. Dawes
The British Paper and Board Industry Federation
UK

Peter Cutler
Wiggins Teape
UK

Derrick Croxon
European Tissue Symposium
UK

M.J. Gadd
Read Paper & Board (UK) Ltd
UK

Martin Glass
ECC International Ltd.
UK

Mike Shaylor
Bowater Scott Ltd.
UK

John Veness
ECC International Ltd.
UK

Agustin Aishemberg
Fabrica Nacional de Papel S.A.
Uruguay

Juan Carlos Benech
Fabrica Nacional de Papel S.A.
Uruguay

Norma Pace (Ms)
American Paper Institute
USA

Robert C. Eisenach
The Mead Corporation
USA

Virginia R. McLain (Ms)
Westvaco Corporation
USA

Allan Whitman
Scott Paper Company
USA

James L. Hutchison
American Paper Institute
USA

Rob Street
Union Camp
USA

R.L. Thompson
Scott Paper Co.
USA

Stephen Larson
Boise Cascade Corporation
USA

Kathryn McAuley (Ms)
Champion International Corporation
USA

John Dodge
Champion International Corporation
USA

Ronald J. Slinn
American Paper Institute
USA

Robert G. Galligan
ITT Rayonier
USA

Steve Whybrew
Weyerhaueser Company
USA

Louis Vargha
Weyerhaeuser Co.
USA

Anibal Valero
APROPACA
Venezuela

FAO Secretariat

Nicos Alexandratos

Mike Arnold

Shelby Campbell Bruzzano (Ms)

Maurizio de Nigris

Carlos D'Ricco

Heiki Huuhtanen

Frederick Keenan

Borje Kyrklund

Leo Letourneau

Leo Lintu

R. Michael Martin

Felice Padovani

Massimo Palmieri

Andreas Polycarpou

Gina Phillips Valeri (Ms)

Philip Wardle

Consultants

William Algar
Australia

Jorge Mascaro
Chile

Markku Simula
University of Helsinki
Finland

Ornella Corridori (Ms)
University of Rome
Italy

Clement Kahuki
Forest Department
Kenya

George Igugu
Federal Department of Forestry
Nigeria

Erik Arvidsson
University of Umea
Sweden

Anders Baudin
University of Umea
Sweden

Lars Lundberg
FIEF
Sweden

Per Thege
Sweden Forest Consulting AB
Sweden

John Adams
DUO (UK) Ltd.
UK

Michael Galsworthy
Hawkins Associates
UK

Richard Harris
BIS Marketing Research Ltd.
UK

Christopher Upton
University of Reading
UK

Joseph Buongiorno
University of Wisconsin
USA

Robert Coen
McCann-Erickson Inc.
USA

Roger Hurwitz
Massachusetts Institute of Technology
USA

Emil Jones
USA

John Kalish
USA

John Koning
US Forest Service
USA

Daniel Navon
US Forest Service
USA

Peter Oliver
Andover International Associates
USA

Allen Whitman
Whitman Associates Inc.
USA

Richard Young
Chase Econometrics
USA

ANNEX III

List of Meetings

FAO ADVISORY COMMITTEE ON PULP AND PAPER, 13rd Session - FAO, Rome, 9-11 June 1982

First Subgroup Meeting - FAO, Rome, 4-5 October 1982

Second Subgroup Meeting - FAO, Rome, 8-9 February 1983

INDUSTRY WORKING PARTY (IWP) MEETING ON DATA - FAO, Rome, 4-6 May 1983

Third Subgroup Meeting - FAO, Rome, 7 June 1983

FAO ADVISORY COMMITTEE ON PULP AND PAPER, 14th Session, FAO, Rome, 8-10 June 1983

IWP MEETING ON MODEL STRUCTURE AND ASSUMPTIONS - American Paper Institute, New York, 5-7 October 1983

IWP MEETING ON MODEL STRUCTURE AND ASSUMPTIONS - Australian Paper Manufacturers Ltd., Melbourn, 16-17 October 1983

Fourth Subgroup Meeting - FAO, Rome, 16 January 1984

IWP MEETING ON ECONOMIC INDICATORS - FAO, Rome, 8-9 March 1984

IWP MEETING ON NEWSPRINT - Canadian Pulp and Paper Association, Montreal, 25-27 April 1984

Fifth Subgroup Meeting - FAO, Rome, 15 May 1984

FAO ADVISORY COMMITTEE ON PULP AND PAPER, 15th Session, FAO, Rome, 16-18 May 1984

IWP MEETING ON WRAPPING AND PACKAGING PAPER - Swedish Pulp and Paper Association, Stockholm, September, 1984

IWP MEETING ON PRINTING AND WRITING PAPER - FINNPAP, Helsinki, October 1984

IWP MEETING ON HOUSEHOLD AND SANITARY PAPER - Bowater Scott Ltd., London, 12-13 November 1984

INDUSTRY WORKING GROUP MEETING ON CONTAINERBOARD - American Paper Institute, New York, 18-19 December 1984

INDUSTRY WORKING GROUP MEETING ON FOLDING BOXBOARD - SCANPAPP, Copenhagen, 14 February 1985

Sixth Subgroup Meeting, FAO, Rome, 21 February 1985

IWP REGIONAL MEETING FOR LATIN AMERICA - CICEPLA, Santiago, Chile, 11-12 April 1985

IWP REGIONAL MEETING FOR ASIA AND THE PACIFIC - Japan Paper Association, Tokyo, Japan, 15-16 April 1985

IWP REGIONAL MEETING FOR EUROPE AND NORTH AMERICA - FAO, Rome, 2-3 May 1985

Seventh Subgroup Meeting, FAO, Rome, 11 June 1985

70

FAO ADVISORY COMMITTEE ON PULP AND PAPER, 16th Session, - FAO, Rome, 12-14 June 1985

INDUSTRY WORKING GROUP MEETING ON PERSPECTIVES FOR WRAPPING AND PACKAGING PAPERS - Verband Deutscher Papierfabriken (VDP), Bonn, 25-26 September 1985

IWP MEETING MEETING ON FIBRE INPUT - American Paper Institute, New York, 21-22 November 1985

IWP MEETING ON SUPPLY - Associação Nacional dos Fabricantes de Papel e Celulose, Sao Paulo, 26-27 November 1985

IWP MEETING ON DRAFT REPORT - FAO, Rome, 17-19 March 1986

ANNEX IV

References and Background Documents

Full detail of the projections are published in *Forest Products World Outlook Projections*, FAO, 1986.
A selection of background documentation to the preparation of the *FAO Outlook StudY* is published in *Readings from the Outlook Study for Supply and Demand of Pulp and Paper* FAO, 1986.
The following is a complete list of reports prepared during the study with main references:

Adams, Darius M.   May 1983
MODELLING WORLD TRADE TROPICAL TIMBERS

American Paper Institute   November 1984
RECENT MARKET PRESSURES IN PLASTICS PACKAGING

Baudin, Anders and Lundberg, Lars   September 1984
DEMAND MODELS FOR NEWSPRINT - PRELIMINARY RESULTS AND PROJECTIONS

Baudin, Anders and Lundberg, Lars   1984
DEMAND MODELS FOR MECHANICAL WOOD PRODUCTS

Baudin, Anders and Lundberg, Lars   April 1984
DEMAND FOR NEWSPRINT - ECONOMETRIC MODELS AND PROJECTIONS

Baudin, Anders and Lundberg, Lars   August 1983
ANALYSIS OF THE DEMAND FOR FOREST PRODUCTS - PRELIMINARY SURVEY

Baudin, Anders and Lundberg, Lars   August 1984
DEMAND MODELS FOR PRINTING AND WRITING PAPER

Baudin, Anders and Lundberg, Lars   June 1984
PROJECTIONS OF DEMAND FOR SAWNWOOD AND WOOD BASED PANELS,
1985-2000 - WORKING REPORT

Baudin, Anders and Lundberg, Lars   November 1984
A DEMAND MODEL FOR HOUSEHOLD AND SANITARY PAPER - RESULTS AND
PROJECTIONS - WORKING REPORT

Baudin, Anders and Lundberg, Lars   October 1984
DEMAND MODELS FOR COATED AND UNCOATED PRINTING AND WRITING PAPER
- ADDITIONAL RESULTS AND PROJECTIONS

Baudin, Anders and Lundberg, Lars   September 1984
DEMAND MODELS FOR OTHER PAPER AND PAPERBOARD - PRELIMINARY
RESULTS AND PROJECTIONS

Baudin, Anders and Lundberg, Lars   March 1985
DEMAND FOR PAPER AND PAPERBOARD
In Product Detail OECD Countries

Baudin, Anders   May 1985
UNCERTAINTY OF DEMAND PROJECTIONS BASED ON POOLED TIME SERIES,
CROSS SECTION MODELS

Bergstedt, Stig   November 1984
PACKAGING IN SWEDEN

Buongiorno, Joseph   May 1985
STABILITY OF INCOME AND PRICE ELASTICITIES IN THE DEMAND FOR FOREST
PRODUCTS

Canadian Pulp and Paper Association   October 1984
FIBRE CONSUMPTION IN THE CANADIAN PAPER AND BOARD INDUSTRY

Chase Econometrics   1986
LONG TERM REPORT

Coen, Robert J.   October 1984
DEVELOPMENTS IN ADVERTISING AND IMPLICATIONS FOR COMMUNICATIONS
MEDIA

D'Ricco, Carlos   October 1985
FURNISH COMPOSITION
Industry Data on Fibre Input for Paper Output

European Tissue Symposium   December 1985
DEVELOPMENT OF HOUSEHOLD AND SANITARY PAPERS IN WESTERN EUROPE
TO 1995

FAO   April 1982
23rd ADVISORY COMMITTEE NOTE ON PULP AND PAPER OUTLOOK STUDIES

FAO   April 1983
PULP, PAPER AND PAPERBOARD TOTAL CAPACITY SURVEYS 1968-1983

FAO   April 1983
WASTE PAPER DATA 1961-1982

FAO   June 1983
24th ADVISORY COMMITTEE, 1st, 2nd, 3rd SUBGROUP REPORTS, IWP REPORT -
MEETING ON DATA - ROME

FAO   October 1983
IWP REPORTS - MEETINGS ON MODEL STRUCTURE AND ASSUMPTIONS NEW
YORK AND MELBOURNE

FAO   January 1984
SUBGROUP REPORT - FOURTH MEETING - ROME

FAO   March, 1984
FAO COMPENDIUM OF MACRO-ECONOMIC INDICATORS 1961-2000

FAO   March 1984
IWP REPORT - MEETING ON ECONOMIC INDICATORS - ROME

FAO   April 1984
IWP REPORT - MEETING ON NEWSPRINT - MONTREAL

FAO   May 1984
SUBGROUP REPORT - FIFTH MEETING - ROME

FAO   August 1984
THE PULP AND PAPER INDUSTRY DATA IN THE OECD 1961-1981

FAO   September 1984
IWP REPORT - MEETING ON WRAPPING AND PACKAGING PAPER - STOCKHOLM

FAO	October 1984
IWP MEETING ON PRINTING AND WRITING PAPER - HELSINKI	
FAO	November 1984
IWP MEETING ON HOUSEHOLD AND SANITARY PAPER - LONDON	
FAO	January 1985
DEMAND FOR NEWSPRINT	
FAO	February 1985
SUBGROUP REPORT - SIXTH MEETING - Rome	
FAO	March 1985
PROJECTIONS OF DEMAND FOR PAPER AND PAPERBOARD	
FAO	May 1985
IWP REGIONAL MEETINGS FOR LATIN AMERICA, ASIA AND THE PACIFIC, AND EUROPE AND NORTH AMERICA Santiago, Tokyo, Rome	
FAO	May 1985
OUTLOOK STUDY FOR SUPPLY AND DEMAND OF PULP AND PAPER - DRAFT REPORT	
FAO	June 1985
PULP AND PAPER CAPACITIES 1984-89	
FAO	June 1985
SUBGROUP REPORT - SEVENTH MEETING - Rome	
FAO	November 1985
IWP REPORT - MEETING ON FIBRE INPUT NEW YORK	
FAO	November 1985
IWP REPORT - MEETING ON SUPPLY SAO PAULO, Brazil	
FAO	February 1986

1984 Yearbook of forest products

Galsworthy, Michael J.	August 1984
FILLERS COATINGS AND PIGMENTS	
Glass, Martin and Veness, J.C.	October 1984
OUTLOOK FOR PRINTING AND WRITING PAPERS IN WESTERN EUROPE TO 1988	
Graff, Peter	March 1984
NEWSPRINT - TV'S BELATED VICTIM?	
Graff, Peter	May 1983
PAPER CONSUMPTION - A GLOBAL FORECAST	
Graff, Peter	November 1984
PRINTING PAPER - SUBJECT TO TV'S CLEMENCY?	
Harris, Richard	April 1984
IMPACT OF COMMUNICATIONS MEDIA DEVELOPMENTS ON CONSUMER PUBLICATIONS AND BUSINESS DOCUMENTATION	

74

Hurwitz, Roger
TRACKING INFORMATION FLOWS AND MEDIA TRENDS
April 1984

Hurwitz, Roger
FROM INFORMATION FLOWS TO MEDIA TRENDS
May 1984

Japan Paper Association
LONG-TERM FORECAST FOR JAPANESE PAPER AND PAPERBOARD
January 1984

Japan Paper Association
REVISED LONG-TERM FORECAST FOR JAPANESE PAPER AND PAPERBOARD
INDUSTRY
January 1984

Jerkeman, Per
PACKAGING MATERIALS OF THE FUTURE — PAPER OR PLASTIC?
September 1984

Lintu, Leo
NOTE ON FAO WASTE PAPER SURVEYS
1983

Lundberg, Lars
DEMAND MODELS FOR PAPER AND PAPERBOARD USED IN THE FAO OUTLOOK
STUDY
April 1985

Lundberg, Lars
DEMAND MODELS FOR PAPER AND PAPERBOARD USED IN THE FAO OUTLOOK
STUDY
May 1985

Martin, R. Michael
VALIDATION OF FAO PULP AND PAPER CAPACITIES
September 1985

Martin, R. Michael
FAO ANALYSIS OF THE SUPPLY OF PAPER PRODUCTS
October 1985

McAuley, Kathryn F.
INTERNATIONAL MACRO-ECONOMIC FORECASTS AVAILABLE FROM US
CONSULTING FIRMS
October 1983

McAuley, Kathryn F.
INTERNATIONAL MACRO ECONOMIC FORECASTS AVAILABLE FROM U.S.
ECONOMIC CONSULTING FIRMS — UPDATE
March 1984

Meister, Irene
Report prepared by a Subgroup of the Industry Working Party
CONTAINERBOARD
December 1984

Oliver, Peter L.
ELECTRONIC MEDIA AND OTHER PERILS IN FORECASTING WORLD NEWSPRINT
DEMAND
April 1984

Padovani, Felice
FOREST DATABASE DESIGN 1983
1983

Soetikno, Abubakar
THE ASEAN PULP AND PAPER INDUSTRY
October 1983

Teras, Timo
PULP AND PAPER SUPPLY MODEL
April 1984

Thornander, Lars
Report prepared by a Subgroup of the Industry Working Party
FOLDING BOXBOARD
February 1985

United Nations   November 1982
COMPENDIUM OF WORLD DEVELOPMENT INDICATORS 1960-1980

Upton, Christopher   1984
WASTE PAPER DATA 1961-1982

Upton, Christopher   May 1983
NOTES ON FAO DATA SERIES

Wardle, Philip   1983
STRUCTURE AND ASSUMPTION OF THE MODELS TO BE USED IN THE STUDY

Wardle, Philip   October 1984
DERIVED DEMAND FOR INPUT

ANNEX V

List of Tables and Figures

Tables

Figures